Taking Sustainable Cities Seriously

American and Comparative Environmental Policy
Sheldon Kamieniecki and Michael E. Kraft, editors

Russell J. Dalton, Paula Garb, Nicholas P. Lovrich, John C. Pierce, and John M. Whiteley, *Critical Masses: Citizens, Nuclear Weapons Production, and Environmental Destruction in the United States and Russia*

Daniel A. Mazmanian and Michael E. Kraft, editors, *Toward Sustainable Communities: Transition and Transformations in Environmental Policy*

Elizabeth R. DeSombre, *Domestic Sources of International Environmental Policy: Industry, Environmentalists, and U.S. Power*

Kate O'Neill, *Waste Trading among Rich Nations: Building a New Theory of Environmental Regulation*

Joachim Blatter and Helen Ingram, editors, *Reflections on Water: New Approaches to Transboundary Conflicts and Cooperation*

Paul F. Steinberg, *Environmental Leadership in Developing Countries: Transnational Relations and Biodiversity Policy in Costa Rica and Bolivia*

Uday Desai, editor, *Environmental Politics and Policy in Industrialized Countries*

Kent E. Portney, *Taking Sustainable Cities Seriously: Economic Development, the Environment, and Quality of Life in American Cities*

Taking Sustainable Cities Seriously

Economic Development, the Environment, and Quality of Life in American Cities

Kent E. Portney

The MIT Press
Cambridge, Massachusetts
London, England

This book was set in Sabon by SNP Best-set Typesetter Ltd., Hong Kong Printed on recycled paper and bound in the United States of America.

Library of Congress Cataloging-in-Publication Data

Portney, Kent E.
 Taking sustainable cities seriously : economic development, the
 environment, and quality of life in American cities / Kent E. Portney.
   p. cm.—(American and comparative environmental policy)
 Includes bibliographical references and index.
 ISBN 0-262-16213-X (hc. : alk. paper)—ISBN 0-262-66132-2
 (pbk. : alk. paper)
   1. Urban ecology—United States. 2. Urban policy—United States.
 3. Sustainable development—United States. I. Title. II. Series.
 HT243.U6 P67 2003
 307.76'0937—dc21   2002071784

10 9 8 7 6 5 4 3 2

Contents

Series Foreword   vii

Preface   ix

Acknowledgments   xiii

1 Sustainability, Sustainable Economic Development, and
   Sustainable Communities: The Conceptual Foundations of
   Sustainable Cities   1

2 Measuring the Seriousness of Sustainable Communities   31

3 The Environment, Energy, and Sustainable Cities   77

4 The Economic Development Side of Sustainability:
   Growth versus Smart Growth   101

5 The Communitarian Foundations of Sustainable Cities:
   A Solution to the Tragedy of the Commons, the NIMBY
   Syndrome, and Transboundary Impacts?   125

6 Is a Sustainable City a More Egalitarian Place? Sustainable
   Communities, Environmental Equity, and Social Justice   157

7 Cities That Take Sustainability Seriously? A Profile of Eight
   Cities   177

8 Sustainable Cities in Practice: More Cities, More Questions   221

References   251

Index   271

Series Foreword

Thirty years ago, states and localities looked to the national government to design and implement policies for control of air and water pollution, management of hazardous waste, and a multitude of other environmental problems. The system of federally dominant environmental regulation continues in the early twenty-first century. Today, however, it is subject to more criticism and inspires a continuing search for alternatives that better suit the environmental issues faced today—from pollution control to population growth, the consequences of economic development, land use, energy use, and transportation. One of those alternatives is an attempt to integrate environmental, economic, and social concerns under the general banner of sustainability. Some of the most creative and important of these efforts have occurred at the local and regional level, not only in the United States but in Canada, Europe, and elsewhere around the world.

In this book, Kent Portney examines efforts to promote sustainability from several new perspectives. He argues that a logical but neglected focus of inquiry has been what urban areas—cities, not just small towns and rural communities—have been doing to consider and act on sustainability issues. He also goes beyond the single case-study approach so common in the study of sustainability to ask more systematically why some cities embark on sustainability initiatives (that is, why they take sustainability seriously) and others do not, and what factors are most responsible for the development and operation of sustainability programs. He focuses on a relatively small number of American cities that can be characterized as true innovators in their efforts to become more sustainable. Moreover, he wants to know what difference these policies and programs are making to the cities that have adopted them. Are those cities better off or not? Which among the varied sustainability actions and programs have proved to be the most effective?

It is not easy to answer the kinds of questions that Portney addresses in this book, but it is important to try. We need to advance the study of sustainability beyond what scholars have been able to offer to date. Previous work has consisted in large part of descriptive accounts of community sustainability actions, the problems they have encountered, and some of the lessons that might be drawn from specific cases. Portney's comparative analysis moves us much closer to understanding the correlates, causes, and consequences of sustainability initiatives. He also sets out an intriguing agenda for future research that can expand our knowledge of sustainability initiatives in the years ahead if others take up his challenge to frame research issues in a more theoretically rich manner.

Portney's work illustrates the kind of books published in the MIT Press series in American and Comparative Environmental Policy. We encourage books that examine a broad range of environmental policy issues. We are particularly interested in volumes that incorporate interdisciplinary research and focus on the linkages between public policy and environmental problems and issues both within the United States and in cross-national settings. We anticipate that future contributions will analyze the policy dimensions of relationships between humans and the environment from either an empirical or theoretical perspective. At a time when environmental policies are increasingly seen as controversial and new approaches are being implemented widely, the series seeks to assess policy successes and failures, evaluate new institutional arrangements and policy tools, and clarify new directions for environmental politics and policy. These volumes will be written for a wide audience that includes academics, policymakers, environmental scientists and professionals, business and labor leaders, environmental activists, and students concerned with environmental issues. We hope that these books contribute to public understanding of the most important environmental problems, issues, and policies that society now faces and with which it must deal well into the new century.

Sheldon Kamieniecki, University of Southern California
Michael E. Kraft, University of Wisconsin–Green Bay
American and Comparative Environmental Policy Series Editors

Preface

Today, there are at least twenty-five major cities in the United States that have invested significant amounts of time, resources, and political capital in the development of initiatives to pursue some form of sustainability. Cities ranging from Portland and Seattle, San Jose, San Francisco, Santa Monica, and Santa Barbara, Boulder, Scottsdale, Tucson and Phoenix, Austin, Chattanooga, Jacksonville, Tampa, and Orlando, Indianapolis and Milwaukee, Boston, Cambridge, and Brookline and many others have started to take the idea of sustainability seriously as a matter of public policy. Numerous other cities are contemplating creating such programs, and cities in Canada and Europe have been world leaders in promoting the idea of sustainability within the context of local governments.

Given this fact, it is perhaps surprising that so little serious, hypothesis-driven research has been conducted to examine these cities' programs and initiatives. There are numerous case studies of cities' experiences, usually with an eye toward providing some level of detail concerning the "best practices" of sustainable cities that can presumably be copied or emulated elsewhere. Sometimes case studies try to relate the experiences of cities as they embarked on a sustainability initiative. Sometimes they try to identify problems or impediments that were encountered in a specific city so that those problems can be avoided by other cities. Sometimes they are prepared for the purpose of providing direct guidance to planners or others who wish to initiate a sustainability effort. There are numerous works that advocate for the creation of sustainable cities, providing either a conceptual rationale for them, or "how-to" manuals for those involved in their development.

Generally missing from the literature on sustainable cities or sustainable communities are systematic comparisons of cities conducted for the purpose of examining specific research hypotheses. There are very few research efforts directed toward examining hypothetical correlates, causes, or consequences, of sustainability efforts in cities. Even though many case studies carry with them implicit or explicit messages about why they worked or why they didn't work, or other lessons that may be grounded in some causal conception, there remain many unanswered questions. Why do some cities embark on sustainability initiatives and others not? What factors are most conducive to the creation and operation of sustainability efforts? Are cities better off when they do establish and implement such programs? What are the most effective sustainability program elements or components? These are a few general questions of the sort that could be addressed.

By and large, these kinds of questions have all but been ignored in the literature of sustainable cities. Perhaps more germane to this book, the vast majority of the literature on sustainable cities and communities does not even frame the central questions in a form that can be addressed in a systematic way. Existing studies rarely discuss the theoretical foundations of cities' efforts, and the lack of comparisons inevitably limits the kinds of inferences that can be drawn from them. From a theoretical point of view, why would one expect some cities rather than others to pursue sustainability? The casual observer can easily conjure up some plausible hypotheses, but the published literature is largely absent of discussions that frame issues in this way. Is it easier for a smaller city or for a larger city to embark on a serious sustainability effort? A case study of the highly successful effort in Boulder might lead one to the conclusion that smallness is an advantage. But a case study of Seattle, or Portland, or San Jose might lead one to the conclusion that largeness and economies of scale present significant advantages. Systematic comparisons across cities reported here suggest that the size of the city makes little difference. Much the same can be said of numerous other characteristics of cities. In short, until the study of sustainable cities can move toward making such systematic comparisons, it will be difficult to know which plausible hypotheses seem to be correct, or the conditions under which they seem to hold.

This book represents a first step in redressing this omission. Sustainability initiatives in cities promise to make significant differences in the quality of life and the quality of the environment for their residents. They also promise a vehicle for promoting ecosystem health and protecting the environment worldwide. Yet if this promise has any hope of being fulfilled, our understanding of what works and what doesn't work, of what people will accept and what they won't accept, and what conditions contribute to more effective programs or initiatives, must advance significantly. If sustainable cities are to become more than a fad, more than a passing fancy pursued by the eccentric few, then much more research must be conducted.

This book also sets the stage for extensive future empirical research by providing discussions of the conceptual linkages between sustainability in theory and in cities' practice and beginning the process of framing many of the issues surrounding sustainable cities in ways that will facilitate the specification of causal hypotheses. Also, this process begins by extracting debatable, sometimes controversial, testable social scientific hypotheses from the existing and extant literatures on sustainability, sustainable communities, or sustainable cities, or from the experiences of cities as they endeavor to move toward becoming more sustainable. Sometimes it provides basic information to address these hypotheses, and sometimes it does not. It provides some of the important conceptual underpinnings for future empirical analysis, as found in the case of the discussion of operationalizing the concept of "taking sustainable cities seriously." It does not seek to be as rigorous as future research will need to be, but it does attempt to provide the foundations for greater rigor in that research. The hope that emerges from this book is that advocates of sustainable cities will take a hard, critical look at what has been tried so far, will seek an even greater understanding of how their concepts and practices can be improved, and will adopt an even more empirically grounded validation of what they wish to accomplish.

This book is motivated by my belief that substantial, if not fundamental, changes need to be made in American society if my children and their children are going to have the opportunity to thrive. I grew up in an era when virtually no one paid attention to what we were doing to the earth. I grew up in a town in New Jersey called Glassboro, next to the town of Pitman, which had perhaps the worst hazardous waste site in the United States. The site, now known as the Lipari Landfill site, lay adjacent to a small stream and lake right near the border of these two towns. The array of environmental hazards present in that area were unbeknownst to nearly everyone there, and I shudder to think what kinds of environmental toxins I was exposed to throughout my childhood. While the lake was legally closed because of its polluted status, no one ever told the residents, and recreational uses of the area by me, my friends, and hundreds of others were frequent. If this were not bad enough, opportunities for other routine exposures to environmental contaminants were numerous. I still remember when I was perhaps seven or eight years old, my friends and I would ride our bicycles behind the truck that was spraying DDT fog for mosquito control. As a high schooler, with no inkling of the risks involved, I worked as an installer of asbestos siding on houses.

We have seen a great deal of change since that time. Due in large part to aggressive environmental protection efforts at the federal and state levels, it has become much more difficult to create toxic sites such as the Lipari Landfill. Opportunities for exposures to environmental contaminants have largely decreased, although certainly not disappeared. Slowly, our collective consciousness has changed so that we are more attuned to the connection between the environment and human health than we were when I was young. I am guardedly optimistic about what can be accomplished through public policy and through carefully defining and designing roles for government. I know and understand the social, economic and political forces that are at work to undermine what can be accomplished. Ultimately, I am optimistic about the prospects of transforming our nation and the world into sustainable places. My interest in sustainable cities reflects this optimism and is dedicated to finding realistic and practical ways to make such transformations happen.

Acknowledgments

My first instinct in making acknowledgments is for me to consider how I have come to know and understand the materials in this book. Who are the people who have influenced me to think about these issues in this particular way? There is certainly no one but me to blame for this book, but there are a great number of people who have been instrumental in helping me develop many of my ideas toward sustainability and sustainable cities. There are three people in particular who have, over the years, contributed a great deal to how I view sustainable cities. Dan Mazmanian, C. Erwin and Ione L. Piper Dean and Professor of the School of Policy, Planning and Development at the University of Southern California, and Michael Kraft, Professor of Political Science and Herbert Fisk Johnson Professor of Environmental Studies, University of Wisconsin–Green Bay, are two of these people. Their collective works have helped to define what environmental policy research is about today in political science and their co-authored book *Toward Sustainable Communities: Transition and Transformations in Environmental Policy* has helped to elevate the level of discourse about such issues. They have provided a fresh look at sustainable communities issues, have each had a major influence on my thinking about and approaches to sustainability. The third person, Elizabeth Kline, has been about as tireless as anyone I know in dedication to the pursuit of sustainable communities. She has always brought clarity of thought and conceptualization to her convictions about sustainable communities issues, and as a result has made me rethink my views many times over. Learning about her experiences in a number of cities' initiatives gave me a foundation from which to work, without which I would have had great difficulty conducting this research.

I would be greatly remiss if I did not mention by name all of the students who enrolled in my undergraduate seminars on sustainable communities at Tufts, and who provided me with a forum in which to test my ideas. I feel very fortunate to have been able to try out my ideas in the presence of Lindsay Barton, Jessica Boschee, Jackie Brot, Tom Damassa, David Drucker, Eric Greenberg, Katharine Hand, Josh Jarvis, Sally Mimms, Jesse Montero, Sara Morency, Ryan Murray, Johanna Neumann, Jeanneatte Park, Jen Ritter, Gerard "Jay" Rooney, and Frances Switkes, all of whom participated in my seminars, collected some data, conducted some interviews, and otherwise helped me understand what was going on in specific cities. Also at Tufts, I am very fortunate to have colleagues who are committed to creating an intellectually engaging atmosphere and who routinely read my work and provide me with constructive feedback. My colleague in political science, Jeff Berry, and my colleague in the Department of Urban and Environmental Policy, Julian Agyeman, read the manuscript and provided invaluable insight into how it could be improved, not all of which I was able to take advantage of.

At MIT Press, two people deserve special mention. I owe great gratitude to Clay Morgan, who was supportive and helpful in the production of this book from the first moment I discussed it with him. Deborah Cantor-Adams production edited the book with great care. I would also like to thank Ann Jonas, Sara Meirowitz, Laurel Ibey, and Chryseis Fox for their assistance. Together, they made the process of preparing this book a delightful and rewarding experience. I would also like to thank the American and Comparative Environmental Policy Series co-editors, Michael Kraft and Sheldon Kamieniecki, for their support, encouragement, and constructive criticism. Of course, despite all the contributions of these many fine people, any errors of omission or commission are my own.

Taking Sustainable Cities Seriously

1

Sustainability, Sustainable Economic Development, and Sustainable Communities: The Conceptual Foundations of Sustainable Cities

What are sustainable cities? That is the essential question that underlies this book. The concept of sustainable communities has blossomed over the last ten years, promising a permanent alteration in the way people see themselves and their communities in relation to their physical and social environments. But before this question can be answered, there are many other related questions that also need to be addressed. For example, what exactly is a sustainable community? What are the conceptual roots of the idea of sustainable communities? Is the concept of a sustainable community applicable to relatively densely populated urban areas, such as a city? Are there cities in the United States that are engaged in efforts to become sustainable? If so, what are these cities actually doing? Are sustainable communities capable of living up to their promise? Is urban sustainability fundamentally different, in some sense, from sustainability applied to other settings? Can sustainability, *writ small*, overcome some of the obstacles confronting more sweeping and geographically ambitious conceptions of sustainability? These are the kinds of questions asked by scholars and practitioners alike, and they are the kinds of questions that animate debate about whether sustainable cities can, and do, work.

The broad concept of sustainability has caught the attention of policymakers and citizens the world over. It has evolved over time, and much of what the term conveys today is considerably different from what it conveyed a decade ago. As the broad concept of sustainability has evolved, so, too, have several of its derivatives—sustainable communities, livable communities, and sustainable cities. Even so, these are not concepts that are susceptible to easy or quick definitions. As Beatley and Manning (1997, 3) point out, "there is a general sense that sustainability is a good thing (and that being unsustainable is a bad thing), but will we know it when we see it?" As appropriate as this question is, there apparently are a number of cities around the United States that feel sanguine enough about what sustainability means that they are willing to try to move forward to achieve it. Without offering a fully developed definition of a sustainable city at this point, suffice it to say here that a sustainable city is a city that is working hard to promote some operational version of sustainability.

This book offers a review and an assessment of sustainable cities in the United States. It starts with general conceptions of sustainability that serve as the underpinnings of sustainable cities, and explicitly links emerging practices to these conceptual underpinnings. It does not attempt to determine whether cities are, or are becoming, sustainable. As will be argued later, the conceptual foundations of sustainable cities inevitably prescribe a very long-term process, perhaps taking decades to achieve substantial results. Are sustainable cities' activities actually making a difference in terms of economic development and environmental quality? As important as this question is, this book does not try to answer it. Since the work toward creating sustainable cities in the United States is, by any yardstick, early in its gestation period, it would be unproductive to try to determine whether this work has succeeded.

However, it is not too early to assess whether cities are, or seem to be, moving down the correct road toward achieving sustainability. Perhaps serving as a sort of intellectual amniocentesis, this book asks the question "how seriously are cities taking the pursuit of sustainability?" In short, any sort of assessment of serious sustainability efforts needs to examine what kinds of activities—policies, programs, organizations, and practices—are being used in U.S. cities in their purported efforts to achieve greater sustainability. If progress toward urban sustainability is to materialize, it must be animated by the activities of people and organizations. This research examines what cities actually seem to be doing in their respective pursuit of sustainability, and compares and contrasts these activities to the kinds of activities that purport to be consistent with various theoretical conceptions of sustainability. Moreover, once there is a clear sense of the range and types of these activities, questions about whether it is reasonable to apply the same yardstick of seriousness across cities, and why some cities seem, at least on the surface, to be taking sustainability more seriously than others, can begin to be addressed. Is it appropriate to judge the seriousness of any two cities' sustainability initiatives by the same criteria and standards? Are there some types of sustainability activities that are taken more seriously than others? Are there some general patterns that suggest reasons why the pursuit of sustainable cities is taken more seriously in some places than in others? If this book does not definitively answer the many questions it raises, at least it begins the process of framing the questions in ways that they can be addressed in future research. However, before these questions and issues are addressed, the concepts of sustainability, sustainable economic development, sustainable communities, and sustainable cities need to be clarified.

The Concepts of Sustainability, Sustainable Development, and Sustainable Communities

The concepts of sustainable cities and sustainable communities have their genetic roots in the general concept of sustainability and its close cousin, sustainable economic development, and, in particular, conceptions of what constitutes a "community." Ever since the term "sustainable communities" was first brought into the lexicon of environmentalism, scholars and practitioners have seized upon it to promote and facilitate various kinds of pro-environmental change. While the term obviously seems to convey great meaning to a wide array of people, the fact is that, as a matter of practice, it has come to mean so many different things to so many different people that it probably does as much to promote confusion and cynicism as positive environmental change. Sustainability is a concept that is fairly abstract and broad, subject to a variety of understandings and meanings. When the concept of sustainability is coupled with the idea of community, which is itself an abstract, and, some would say, almost meaningless concept, finding meaning in the idea of sustainable communities seems hopeless.

But as a matter of practice, the idea of sustainable communities has evolved in such a way as to provide greater meaning than would initially appear. As originally envisioned, the concept of sustainable communities was derived in an attempt to account for a large number and variety of environmental and interpersonal impacts of economic growth, broadly defined, not comfortably accommodated by neoclassical economic theory or practice. In short, sustainable communities have been thought of as mechanisms that can be used to redress the often negative or deleterious environmental and social effects of adherence to mainstream approaches to economic development. In contemporary applications of the concept of sustainable communities, key elements of the original vision are frequently omitted, overlooked, or substantially modified. Before attempting to provide a specific definition of sustainable communities, it is necessary to explore the broader underlying concepts of sustainability.

Sustainability, and its close cousin, sustainable development, are perhaps best thought of as general concepts whose precise definitions have yet to be fully explicated. Charles Kidd argues that there are at least six different historical intellectual strains of thought that underlie the contemporary concept of sustainability, each with its own "slant" or articulation of particularly important foundational issues. He discusses the "ecological/carrying capacity" root the "natural resource/environment" root, the "biosphere" root, the "critique-of-technology" root, and the "ecodevelopment" root (Kidd 1992). Becky J. Brown and colleagues suggest that in contemporary usage, the term *sustainability* has some six different definitions that relate to "sustainable biological resource use," "sustainable agriculture," "carrying capacity," "sustainable energy," "sustainable society and economy," and "sustainable development" (Brown, Hanson, Liverman, and Meredith 1987, 713–719). As will be demonstrated in later chapters, each of these intellectual roots and definitions suggests its own set of yardsticks that could be used to measure how seriously a city takes sustainability, and to some degree each can be found in sustainability efforts across cities. Whether, and the extent to which, a particular city's initiatives are built on the base of any one set or combination of definitions is determined by a variety of local social and political factors. In short, the seriousness of a city's commitment to sustainability is ultimately determined by the nature of the local governance regime.

Without going into great detail about each of these conceptions of sustainability, suffice it to say that there is considerable overlap among them. These concepts and definitions of sustainability have been used to convey many different expressions of environmental priorities, each emphasizing some particular aspect or set of values concerning what it is that should be sustained. Ultimately, they all have, in some direct or indirect way, their primary roots in biology, the biophysical environment, and ecology, particularly in the notion of "ecological carrying capacity." This is certainly true of what Kidd calls the "ecological/carrying capacity" intellectual root of the concept of sustainability, and what Brown et al. call "carrying capacity" definition, but it is also deeply embedded in other definitions as well.

Carrying capacity focuses on the idea that the earth's resources and environment have a finite ability to sustain or carry life, particularly animal life. Similarly, a particular ecosystem has a finite ability to sustain the life contained there. When the demands move beyond the carrying capacity of the earth or of a particular ecosystem; i.e., when populations of animals exceed the capacity to support them, species collapse will occur. A central problem concerning the earth's carrying capacity is the fact that population growth itself inevitably leads, in a Malthusian sense, to increasing scarcity of the very resources needed to sustain life, including life of humans. Efforts have been made to distinguish the "maximum carrying capacity" and the "optimum carrying capacity," where the former refers to the largest population size that, while theoretical sustainable, would place the earth at a threshold that is vulnerable to even small changes in the environment. The latter is defined as a smaller, more desirable population size that is less vulnerable to environmental disruptions (Odum 1983). Of course, not everyone accepts the notion that the earth's carrying capacity is indeed finite. As is discussed in greater detail later, optimists suggest that technology makes possible, and even likely, expansion of the earth's carrying capacity.

Coupled with the notion of a limited carrying capacity is the idea that human activity, as currently practiced, is largely unsustainable. In other words, most of human activity depletes rather than replenishes or sustains the earth's resources that contain the capacity to carry life. When people engage in rational economic behavior, they contribute to the depletion of those resources. Markets, as the argument goes, more often than not, create incentives for resource depletion, and thereby undermine the earth's carrying capacity. Nowhere is this more true than in areas involving "commons" resources, including much of the earth's air and water. But human behaviors, whether market-driven or not, often contribute to diminishing carrying capacity of the earth. Even for those who believe that technology can intervene, there is concern that the net balance between what technology can do to enhancing the earth's carrying capacity is more than offset by humans' abilities to deplete it.

This idea of sustaining carrying capacity applies most directly to issues involving renewable resources, such as forests, ocean fisheries, and perhaps the use of soil. In this context, efforts are made to define uses of these resources such that they are not depleted faster than they can be replenished. For example, in an ocean fishery, efforts are made to limit the quantity of fish removed to an amount that does not exceed the ability of the fish to reproduce, maintain, or even grow their populations. Efforts to achieve at least some sort of steady state are examples of efforts to achieve environmental sustainability. Sustainability efforts are not, in principle, particularly new. Contemporary agricultural practices evolved to become more sustainable in the United States since the days of the "dust bowls" during the 1920s and 1930s, when severe drought conditions coupled with poor agricultural practices depleted the soil and decimated the farm industry. Today, conceptions of sustainability are quite common across a wide array of renewable resource types.

Sustainability, then, is most frequently associated with maintaining the earth's carrying capacity, usually through alteration of individual and collective human behavior. Behaving in ways that reduce the rate of population growth, and that find alternatives to depleting natural resources, is certainly consistent with the idea of sustainability. In terms of human behavior, what may be required to maintain the earth's carrying capacity is not well understood or agreed upon, and may in fact be inconsistent with basic values that are prevalent in the industrialized and industrializing worlds. In arguing that sustainability is as much an ethical principle as a set of environmental results, Robinson et al. suggest that "sustainability is defined as the persistence over an apparently indefinite future of certain necessary and desired characteristics of the sociopolitical system and its natural environment" (Robinson, Francis, Legge, and Lerner 1990, 39). What this means is that maintaining the earth's carrying capacity is in large part a function of the social and political values that define and prescribe human behaviors. Achieving sustainability, then, according to this line of reasoning, apparently requires some types of sociopolitical characteristics and values rather than others. This is an issue that will be addressed in later chapters.

The idea of *sustainable development* bears a close relationship to that of *sustainability*, although as the term is used, it brings elements of economic activity more explicitly into the equation. Sustainable development, or its sibling—sustainable *economic* development—a frequent variant, reflects concern for the ways that nations develop and grow their economies. This concern suggests that pursuit of maximum and rapid economic growth, as it is traditionally defined, particularly for developing countries, places extreme burdens on the ecological carrying capacity of the earth. To be sure, over the last decade the concept of sustainable development has provided the foundations for the idea that the pursuit of economic growth must be accompanied by significant consideration to the ecological impacts.

The concept of sustainable development, to a large degree, shifts the emphasis away from mere concern about the environment to include explicit concern about economic development. The argument often put forth is that the wrong kind of economic development will not only deplete the earth's resources and damage the earth's ecological carrying capacity, in the long run it will also undermine achievement of economic growth itself. Unsustainable economic development is just as much about being unable to sustain economic growth as it is about exceeding the earth's ecological carrying capacity.

The linkage between sustainability and economic development *writ large* began to emerge as central issues through the 1970s, when a number of international development programs, including those operated by, or with the assistance of, the World Bank, the International Monetary Fund (IMF), and U.S. Agency for International Development (USAID) came under fire for using their extensive financial resources to inadvertently promote environmental degradation under the guise of economic development in Third World and developing nations. Many nongovernmental organizations took great issue with these development programs, suggesting that they needed to become much more sensitive to the indigenous environments and peoples where their financial resources were being used (Fox and Brown 1998). By the late 1970s, the idea of pursuing this kind of environmentally sensitive economic growth, or *ecodevelopment*, as it was termed by Ignacy Sachs, had found its way into the works of the United Nations Environmental Programme (Kidd 1992, 18).

Sustainable development achieved elevated recognition and legitimacy in the late 1980s, when in 1987 the United Nations' World Commission on Environment and Development (WCED), also commonly known as the Brundtland Commission after its Chair, former Norwegian Prime Minister Gro Harlem Brundtland, issued its report: *Our Common Future*. This report was designed to create an international agenda focusing on how to protect the global environment, or as stated in the report, to sustain and expand the environmental resource base of the world. In the process, it put forth the very general notion that sustainable development consists of economic development activity that "... meets the needs of the present without compromising the ability of future generations to meet their own needs" (WCED 1987, 8). Beyond this, the report is rather short on details and specifics. Its definitional contribution clearly comes out of its focus on what might be called cross-generation concerns, and the idea that economic development needs to be viewed over a longer period of time than is usually practiced.

Capturing this cross-generation concern in the U.S. context, the National Commission on the Environment (NCE) put forth a similar set of conceptual definitions. The 1993 report of this Commission suggested the need for the United States to pursue a:

strategy for improving the quality of life while preserving the environmental potential for the future, of living off interest rather than consuming natural capital. Sustainable development mandates that the present generation must not narrow the choices of future generations but must strive to expand them by passing on an environment and an accumulation of resources that will allow its children to live at least as well as, and preferably better than, people today. Sustainable development is premised on living within the earth's means. (NCE 1993, 2)

These discussions of sustainable development provide a basic conceptual framework for organizing thinking about sustainability, but of course there are many questions left unanswered—questions whose answers are critical for formulating specific applications or measuring results. For example, what exactly is included under the rubric of "natural capital"? In other words, what is it that needs to be sustained?

Is it just natural resources, and if so, which resources? Is it human resources? Is it environmental quality, more broadly defined? Is it ecosystem health? Is it some even more broadly defined quality-of-life? Does it matter who owns the natural capital? Are there necessarily distributional considerations; e.g., does it have to apply to all people? Other questions arise, such as whether sustainability initiatives are really antigrowth. In other words, does advocacy of sustainability really mean the same as promoting no growth? Is it really a position in opposition to economic growth, as commonly defined?

In the conceptual literature, there is a clear sense that sustainability is not antigrowth, per se. Although there is a distinct element of no-growth sentiment in at least one of the intellectual foundations of sustainability (Kidd 1992, 9–12), sustainability is more about the search for peaceful coexistence between economic development and the environment. It is about finding ways to promote growth that are not at the expense of the environment, and that do not undermine future generations. Although much of the conceptual literature on sustainability does not directly address many of these issues, sustainable communities initiatives and practices implicitly provide answers to their underlying questions. The working definitions of sustainability that cities develop themselves provide hints as to what they see as important. In Seattle, sustainability has been defined as "long-term cultural, economic, and environmental health and vitality." In Santa Monica, the sustainable communities initiative seeks ". . . to create the basis for a more sustainable way of life both locally and globally through the safeguarding and enhancing of our resources and by preventing harm to the natural environment and human health." And in Cambridge, Massachusetts, sustainability means the pursuit of ". . . the ability of [the] community to utilize its natural, human, and technological resources to ensure that all members of present and future generations can attain high degrees of health and well-being, economic security, and a say in shaping their future while maintaining the integrity of the ecological systems on which all life and production depends" (Zachary 1995, 8). These working definitions may well provide the foundational frameworks for more elaborate definitions. Indeed, as many cities move through the process of developing some sort of sustainable communities' initiatives, they inevitably provide their answers based on what they believe to be appropriate for themselves. As is discussed later, the answers to these questions are quite varied.

In the context of the global concern for sustainable development of nations, it may seem somewhat incongruous to think of the geographically more narrow idea of sustainable communities. After all, isn't one of the central reasons for the global concern about the environment that small geographic areas are subject to externalities that they cannot control? Yet even in the international context, attention to sustainable development has focused on the local level. When the Brundtland Commission stated that ". . . cities [in industrialized nations] account for a high share of the world's resource use, energy consumption, and environmental pollution" (WCED 1987, 241), it is arguing that serious attention needs to be given to urban sustainability.

The Brundtland Commission report served as the foundation for the discussions and negotiations on sustainable development that took place among nations in the "Earth Summit," held in Rio de Janeiro in June of 1992. One of the results of the Earth Summit was the passage of a resolution often referred to as "Agenda 21," a statement of the basic principles that should guide nations in their quest of economic development in the twenty-first century. As part of the Agenda 21 resolution, significant attention was given to the relationship between national policies and the activities of local governments particularly in chapter 28 of the Agenda 21 resolution. In this section, entitled "Local Authorities' Initiatives in Support of Agenda 21," the link is made clearer. As Agenda 21 states:

Because so many of the problems and solutions being addressed by Agenda 21 have their roots in local activities, the participation and cooperation of local authorities will be a determining factor in fulfilling its objectives. Local authorities construct, operate and maintain economic, social, and environmental infrastructure, oversee planning processes, establish local environmental policies and regulations, and assist in implementing national and subnational environmental policies. As the level of governance closest to the people, they play a vital role in educating, mobilizing, and responding to the public to promote sustainable development. (United Nations Environmental Programme 2000)

Thus, the idea of sustainable communities is born out of an understanding of the importance of individual human behavior, and the local governance context in which that behavior takes place.

The idea of sustainable communities undoubtedly grew out of this particular understanding of the concept of sustainability, one that is grounded in the need to address environmental and livability issues as they affect individual people. But it has also grown out of particular understandings of "community." The concept of community is one that has come to mean so many different things to so many different people that it has been suggested that the term ought to be avoided altogether (Bell and Newby 1974). Community has come to mean everything from neighborhoods, to voluntary organizations, to professional associations, to civic groups, to online Internet chat rooms, and more. In the context of sustainability the idea of community appears to correspond more to geographic areas where problems and issues exist, but it still carries multiple meanings. As a consequence, the idea of sustainable communities itself has come to mean many different things, and encompasses an enormous array of different kinds of activities and types of geographic areas. Sustainable communities are not just about relatively small geographically confined groupings of people. Indeed, as the term has been used, a sustainable community can be anything from a small neighborhood, to a group of people who share some interest, to a program operated by a governmental or nongovernmental organization, to a rather localized ecosystem, to a multistate region encompassing numerous ecosystems.

More typically, however, in common usage the term sustainable communities does seem to encompass or embrace a range of geographically small areas. When the Clinton administration's National Science and Technology Council (NSTC) presented its 1995 National Environmental Technology Strategy, entitled "Bridge to a Sustainable Future," great emphasis was placed on the role of "community" in achieving greater sustainability. Without ever really defining what a community is, the report states that:

Our nation's future strength will in large part be built on the viability of our nation's communities. We must make choices today that increase the sustainability and desirability of our cities, towns, and rural areas if we are to preserve our natural environment and build a strong domestic economy. . . . The largest and most complex class of environmental technologies are those supporting our communities: technologies that transport people or goods, produce and deliver energy, treat water supplies and waste products, provide food, and route and process information. To achieve sustainability, technological solutions must be integrated with the unique economic, social, political, and cultural circumstances of each community. (NSTC 1995, 52, 69–70)

Clearly, the implied meaning of a sustainable community in this report relates to small geographic areas where various new "sustainable" technologies can effectively be integrated with economic, political, and cultural practices that vary from place to place.

In 1999, the U.S. Environmental Protection Agency issued its "Framework for Community-Based Environmental Protection" (CBEP), its version of sustainable communities, which carried a somewhat more explicit statement of what "community" means. This document emphasized a functional, but flexible, definition. According to the EPA,

Intrinsic to CBEP is an understanding of "community." The definition of community endorsed by EPA for CBEP efforts includes places that are associated with an environmental issue(s). The community may be organized around a neighborhood, a town, a city, or a region (such as a watershed, valley, or coastal area). It may be defined by either natural geographic or political boundaries. The key factor is that the people involved have a common interest in protecting an identifiable, shared environment and quality of life. (USEPA 1999, 5)

As one might expect from the EPA's definition, a brief perusal of activities identified under the rubric of sustainable communities turns up a variety of different kinds of activities and related geographic areas, from the clean water initiative in the Fox–Wolf Basin area of Wisconsin, to regional sustainability initiatives in the Great Lakes Basin (Kraft and Mazmanian 1999), to the creation of affordable housing and brownfields development (Beatley and Manning 1997), to eco-industrial parks, to all of the eco-villages, such as EcoVillage at Ithaca (N.Y.) and EcoVillage Cleveland (Gilman and Gilman 1991), eco-neighborhoods (Barton 1998), and coastal zone management initiatives in Virginia (Lachman 1997), to name a few. Suffice it to say that the idea of sustainable communities has come to embrace an enormous array of activities and initiatives. Yet, as the idea has expanded, it has become increasingly difficult to know whether any particular set of activities or initiatives really seem to be consistent with a serious intent to sustain or improve a place's ecological carrying capacity. Much the same issues arise in international conceptions of sustainable communities, which embrace such projects as the Annapurna Conservation Cooperative Project in Nepal, the Pallisa Community Development Trust in Uganda, and the

Second Livestock Project in Mauritania, to name a few (Pye-Smith and Feyerabend 1994).

More recently, and for reasons that are discussed more fully in later chapters, the terminology, if not the underlying concept, adopted in some places focuses on "livable communities" or "livable cities" rather than sustainable cities (Cassidy 1980). Indeed, the livable communities movement and many cities' initiatives predated the sustainable communities movement and cities' initiatives. In most respects, however, the concept of livable communities appears to be virtually the same as its sustainable communities. In 1999, when the Clinton administration announced its new livable communities initiative, the substance of the initiative was very much oriented around sustainability. In *Building Livable Communities: A Report from the Clinton–Gore Administration (1999)*, the initiative is described as having the objective of helping to empower communities to sustain prosperity and expand economic opportunity, to enhance the quality of life, and to build a stronger sense of community, all goals that are very much a part of almost any sustainable communities definition. Partners for Livable Communities (2000) describe such communities in much the same way. The point is, however, regardless of whether the preferred terminology involves that of sustainable communities or livable communities, the substantive differences between the two appear to be slight at best. As is discussed later, there may be important political and strategic reasons for adopting the latter term, but in conceptual terms, they seem to mean about the same thing.

From Sustainable Communities to Sustainable Cities

Partly in order to provide definitional clarity, and to facilitate comparisons (comparing apples to apples rather than apples to oranges), this book focuses not on the "community," however that may be defined, but rather on the city. Although the city may not constitute the most germane geographic unit in terms of ecology or environmental concerns, it is a basic jurisdictional unit in American government. Here, the term *city* is used in a formal sense, referring to the legally defined jurisdicational unit. As a general rule, cities are relatively small divisions of government that nonetheless possess the authority to affect environmental and ecological results. Cities are not coterminous with metropolitan areas except in those few areas around the country in which there has been city–county or metropolitan-wide consolidation. Of course, all cities exist in some larger metropolitan context, and understanding the relationship between the formally defined city and its surrounding must represent an integral part of any examination of specific cities' sustainability efforts.

As diverse as the compositions of U.S. cities are, they are considerably more comparable units than other kinds of communities. Although there may be considerable differences between and among cities in the United States, they share the basic characteristic that they are legally defined entities that have the legitimacy and authority to address issues and problems within their borders. As the Brundtland report states, "local authorities usually have the political power and credibility to take initiatives and to assess and deploy resources in innovative ways reflecting unique local conditions. This gives them the capacity to manage, control, experiment, and lead urban development" for the good of the environment (WCED 1987, 242). Implicit in this statement is the notion that in order to have an impact, government must be involved, and that there needs to be some degree of congruence between the geographic area in which sustainability is to be achieved and the political jurisdiction trying to achieve it. Cities share this important trait. As Agenda 21 noted, ". . . local authorities construct, operate, and maintain economic, social, and environmental infrastructure, oversee planning processes, establish local environmental policies and regulations, and assist in implementing national and subnational environmental policies." Indeed, despite enormous differences, cities share many more characteristics, including a wide array of governmental and policymaking processes, than are typically acknowledged (Waste 1989).

Much has been written about the fact that, in terms of sustainability, ecosystems or species habitats are the appropriate levels at which the environment should be viewed, but in practice there is little correspondence between the geographic area of an ecosystem and the boundaries of governmental jurisdictions. Ecosystems rarely conform to the boundaries of cities or towns, counties, states, election districts, or even nations. This means that no single governmental jurisdiction may possess the authority to deal completely with a particular environmental problem or to achieve sustainability results. Clearly, larger, more encompassing, jurisdictions have advantages in terms of fewer externalities, but there may not be the political will to address sustainability at such higher levels. In the United States, there are many ways in which the federal government could act to work toward greater sustainability, but the contemporary ideological mood, the distribution of power and influence among competing interests in national politics in recent times, and the historical culture of the nation (including the culture of decentralized government and deference to smaller rather than larger units of government, as reflected in U.S. federalism) present significant impediments.

Often, advocates of sustainable development propose reorganizing government to make it conform to environmental needs. Such proposals may take many forms, including efforts at metropolitan consolidation, i.e., the merger of central cities with surrounding suburbs. But it is difficult to imagine wholesale redefinitions of our political jurisdictions to conform to ecosystem boundaries. Rather than making political jurisdictions conform to ecosystems, is it possible to address issues of sustainability within existing political jurisdictions? If the focus on cities as sustainable units represents a concession to political realities, is it, in fact, possible to develop the political will to address issues of sustainability in cities, even if it cannot be done at the state or national level? To the degree this is possible, cities undoubtedly constitute important government jurisdictions.

From the perspective of early twentieth-century political history, progressive advocates once concluded that the power of local business and entrenched political machines precluded the pursuit of their agenda at the city level. This produced an increasing tendency for the progressive agenda to be pushed toward national politics, where progressive interests could find a critical mass of people to mobilize. Perhaps ironically, by the latter part of the twentieth century, the tables had been turned. Two concomitant events have very likely altered the political landscape. In national politics, there has been a secular trend toward election of, and in 1994 control by, conservative Republicans in Congress. Although this has not necessarily foreclosed advocacy opportunities for environmental citizen groups (Berry 1999; Bryner 2000; Shaiko 1999), it has made the pursuit of a sustainability agenda more challenging.

At the same time, U.S. national policy has advocated globalization of the economy, including the development of various free-trade agreements (such as the North American Free Trade Agreement and the Free Trade Agreement of the Americas) that have altered the economies of local communities everywhere. The long-term trend has been toward the decline of manufacturing industries in the United States, and the rise of the service sector. Cities where manufacturing companies once served as the foundation of the local economy now must rely on other sectors to provide needed jobs and employment. Perhaps more important for this discussion, corporations that once dominated local politics and stood in the way of progressive achievements are gone or transformed. Businesses with long-term local ties are frequently no longer locally owned, often now divisions of larger multinational corporations with little or no interest in local politics and policies. Other once powerful locally or family owned businesses have succumbed to the vagaries of economic competition. To the extent that sustainability can be thought of as progressive (Milbrath 1984), potentially this has profound implications for the political feasibility of sustainability in cities.

But, with this said, how does the concept of sustainability specifically apply to such a small geographic area as a city? Is it reasonable or appropriate to even consider using the term *sustainability* in the context of cities and not just in terms of ecosystems, sectors of the economy, or whole nations? Is it appropriate to apply the concept of sustainability to cities in industrialized nations, including those in the United States, when it more clearly applies to the rapidly growing cities in industrializing nations? Certainly, the Brundtland Commission report asserts that urban sustainability is important in industrialized nations if for no other reason than because cities are the places where large and growing proportions of the environmental and social problems reside (WCED 1987, 241–243).

For a variety of reasons, a substantial amount of attention has become focused on the potential for small geographic areas, including cities, to be primary contributors to achieving sustainability. This has given rise to what Marvin and Guy (1998) call a "new localism" of environmental policy. This new localism represents a philosophy that asserts the primacy of local areas and local governments for affecting sustainability (Selman 1996). This philosophy suggests that ordinary people are most likely to pay attention to the physical environment where they see and experience it, and that the governance mechanisms in cities are most likely to be responsive to the environmental concerns of their citizens. Marvin and Guy argue that the tenets of this new localism are essentially flawed, leading to prescriptions for change that they believe are confused and distorted, and ultimately will fail to contribute to increased sustainability. Indeed, there is no shortage of critics and skeptics who dismiss the idea that there can be such a thing as a sustainable city. Perhaps the most optimistic view of the role of cities in achieving sustainability prescribes only a contributory role for local government (Satterthwaite 1997).

But are there solid reasons to believe that cities can constitute efficacious mechanisms for achieving great sensitivity to the health of the environment? Whatever conceptual reasons there might be to suggest that cities can potentially become such mechanisms, these reasons must confront the experience that American cities have long been held to constitute what Harvey Molotch (1976) calls economic "growth machines." In the face of this conception of cities, is it still possible that there can be such an entity as a sustainable city? John Dryzek expands upon the notion that cities and other local jurisdictions offer great opportunities for taking sustainability seriously in his discussion of "radical decentralization" (Dryzek 1987, 216–229). Particularly with respect to the issue of the size of the place that should deal with ecological and environmental issues, he contends that "one central feature of smallness of scale is that a locality both relies upon and has exclusive jurisdiction over the productive, protective, and waste assimilative functions of the ecosystems in its immediate vicinity. . . . Local self-reliance . . . means . . . that communities and their members must pay great attention to the life-support capacities of the ecosystem(s) upon which they rely" (Dryzek 1987, 217–218). Largely because the consequences of not paying attention to the health of the environment are so quickly and completely visible at the local level, Dryzek argues that decentralization of environmental decision-making offers great promise.

Even if decentralization of sustainability activities seems conceptually sound, there is no guarantee that residents of any particular area necessarily are, or will become, serious contributors to achieving specific sustainable results. Are there frameworks that are capable of providing clearer pictures of sustainability—frameworks that would seem to tie localized environmental decision making more directly to issues of sustainability? Advocates of local sustainability have proffered several analogies that help to visualize elements of sustainable cities. For example, some suggest that a sustainable city is one that achieves something of a "closed loop," where human activity in one arena is conditioned on, and conditions, activity in other arenas. Alternatively, imagine a "bubble" being placed over a city or a metropolitan area, and the human activity and its consequences are confronted within this bubble. These analogies essentially imply that what goes on within small geographic areas is indeed relevant to achieving greater sustainability.

Perhaps the single most important analogy to suggest that sustainability at the local level is relevant, even in industrialized nations, comes from the work of Rees (1992; 1996) and Rees and Wackernagel (1994), who developed and applied the idea of "ecological footprint" to urban areas. Ecological footprint refers largely to the size of the environmental impact that is imposed on the earth and its resources. Rees and Wackernagel suggest that the demands that humans place on the earth can be translated into an amount of land necessary to meet those demands. According to them, the average American resident has an ecological footprint that requires about 5 hectares of land (over 12 acres) to provide the shelter, food, and energy to support his/her lifestyle. Just as individual people produce an ecological footprint, so do aggregations of people, including people who live in cities. They suggest that sustainable places seek to purposely reduce and minimize their ecological footprint, i.e., reduce their impacts on the environment. As Beatley and Manning state it, "a sustainable community is a place that seeks to contain the extent of the urban 'footprint' and strives to keep to a minimum the conversion of natural and open lands to urban and developed uses" (Beatley and Manning 1997, 28).

A central element of a sustainable city is how self-sufficient it is. Since the average human apparently needs over 12 acres of land to support his/her levels of consumption, and since no city is so large as to allocate 12 acres per resident, this means that cities are far from being self-sufficient. What this also means is that consumption demands in cities can only be met by drawing on the resources of areas outside of the city. Thus, by definition, cities impose their ecological footprints on areas external to the city, and perhaps in many ways areas outside the city impose ecological footprints inside the city. For example, when a city contracts with a company to dispose of its solid waste, and that company trucks the solid waste outside the boundaries of the city, the city is essentially imposing its footprint on some other place. The residents of Virginia became acutely aware of these externalities when the City of New York contracted to ship much of its solid waste to the south after the City decided to close its own landfill on Staten Island. Of course, the size of the footprint depends on such facts as how much solid waste is produced and transported, what kinds of wastes are transported, and what happens to the waste when it is disposed (is it landfilled, incinerated, etc.)? When residents of areas outside of a city drive their cars to work inside that city, they impose an ecological footprint on the city. The point is that cities are not self-sufficient, and it is difficult to imagine a way to make them so. Rees argues that the lack of correspondence between the geographic area of a city and the geographic area needed to support its population makes the city a less-than-ideal level at which to pursue sustainability. As he warns, " . . . we should remember that cities as presently conceived are incomplete systems, typically occupying less than 1% of the ecosystem area upon which they draw. Should we not be reconsidering how we define city systems, both conceptually and in spatial terms?" (Rees 1997, 308).

As developed in the literature on sustainable communities, this fact does not prescribe that cities should appropriate more land, as many cities do, through annexation or consolidation. This might seem to make a city more self-sufficient because it gives a single set of policymakers legal authority over an area that encompasses the source of many environmental challenges. But of course, it merely expands some aspects of a city's ecological footprint and promotes sprawl, clearly not an optimum use of natural resources (Bank of America 1995; Calthorpe and Fulton 2001; Diamond and Noonan 1996; Dunphy, Brett, Rosenbloom, and Bald 1996; Moe and Wilkie 1997). Presumably, a city's ecological footprint can only be reduced by reducing the amount of land necessary to support that city's consumption and production. Although the common prescription for accomplishing this is to convince people to consume less, there are many other kinds of actions and activities that can be undertaken to reduce the overall ecological footprint of a city. Rees lists five broad areas where city policymakers can work to incrementally reduce the ecological footprints of cities. These include integrated city planning to minimize energy, materials, and land use requirements, increased use of green areas, integrated open space planning, protecting the integrity of local ecosystems, and striving for economic development that has zero net impact on ecosystems (Rees 1997, 308–309). Others have suggested that, while annexation itself may not be the answer to reducing a city's ecological footprint, metropolitan-wide planning and government might be (Calthorpe and Fulton 2001). Of course, defining specific activities and policies that can accomplish these goals is where the real challenge lies. Many of these kinds of activities are discussed later.

Conceptually, the idea that there is an ecological footprint, and that sustainable cities are places that seek to minimize this footprint, makes great sense. It may be difficult to imagine a city that is completely self-sufficient, but making efforts to become more self-sufficient, particularly where the costs of doing so are relatively low, seems almost common-sense. In practice, simply trying to pin down how large any specific city's ecological footprint is, and consequently how it can reduce its size, is no small task. How much solid and hazardous waste comes into a city or leaves a city often cannot be known at any given time. How many goods are imported into and exported from a city is not known in practically any U.S. city. Getting an accurate picture of the environmental impacts of all human activity, including that of people working in the private sector, is almost impossible. As is discussed in chapter 3, however, some cities are making a much more concerted effort to understand the full range of environmental impacts they produce, and work toward reducing those impacts even if the impacts are external to the city itself. If a sustainable city is one that has the smallest possible ecological footprint, then a city that takes sustainability seriously is one that seeks to minimize that footprint. Suffice it to say here that a city that attempts to understand, and subsequently takes steps to reduce, its ecological footprint is more serious about sustainability than one that does not. As a practical matter, many cities that wish to move toward sustainability treat the process somewhat incrementally, focusing on new economic development rather than existing economic activity, trying to make sure that new industries meet a higher standard for the size of their respective ecological footprints. Presumably, the relatively recent trend for cities to define growth boundaries as part of their comprehensive planning represents an effort to pursue economic growth while reducing or minimizing the added increment to their ecological footprints.

The focus on cities is not meant to imply that they are the most appropriate governmental unit to address sustainability issues. Indeed, many issues that sustainability activities and initiatives attempt to address may not be able to be adequately or fully addressed within the context of a single city. But underlying the concept of sustainable communities is the notion that it is possible to make significant strides toward creating healthy and livable places by focusing attention on small geographic areas. In the most sincere terms, sustainable communities put into practice the old adage "think globally, act locally." The prescription to "act locally" is meant, in part, to suggest that advocacy and collective action are important elements, and cities provide a critical mass of people and governmental authority to make collective action possible and effective.

What is perhaps more important than the conceptual argument concerning the appropriateness of focusing on the city is the fact that many cities around the world, including many in the United States, have developed programs that purport to work toward becoming sustainable, that appear to be working toward reducing the size of their ecological footprints or at least their environmental impacts. While the idea of sustainable cities is relatively new to the United States, it is an idea that has taken root much more readily in other parts of the world, particularly Western Europe, Scandanavia (Beatley 2000; Carley and Spapens 1998), and Canada (Pierce and Dale 1999). Based on descriptions of progress in many cities outside of the United States, it would not be a stretch to suggest that the United States lags somewhat behind. Indeed, the International Council for Local Environmental Initiatives, an international organization headquartered in Toronto, established a pilot project called the Local Agenda 21 Model Communities Programme to spearhead sustainable cities efforts in fourteen world cities, including Buga, Columbia; Durban, Johannesburg; and Cape Town, South Africa; Hamilton, New Zealand; Jinja, Uganda; Santos, Brazil; Johnstone Shire, Australia; and Hamilton-Wentworth, Canada, among others (ICLEI 2000). Even in North America, many Canadian cities, including Toronto, may well have accomplished more than any specific city in the United States. No effort is made here to systematically compare U.S. cities with sustainability initiatives to those in other parts of the world. Suffice it to say that the literature seems to be unequivocal that U.S. cities have not achieved the kinds of programs found elsewhere.

Nevertheless, in the U.S. context, cities such as Seattle, Washington; San Francisco, California; Austin, Texas; Chattanooga, Tennessee; Santa Monica, California; Boulder, Colorado; Jacksonville, Florida; Scottsdale, Arizona; and many others profiled here have begun governmental or nonprofit programs to work toward achieving important sustainability results. A reasonably comprehensive list of cities that have developed some sort of sustainability initiative is found in table 1.1. As later discussions demonstrate, most of these programs or initiatives represent fairly comprehensive efforts to improve and protect their cities' environments. Some of the programs are citywide initiatives to address a particular environmental problem. Sometimes they are focused on a particular economic sector or activity (such as household recycling or brownfields development), and sometimes they cut across sectors and activities. Sometimes they operate out of single governmental agencies (an environmental department, a department of public works, a planning department, etc.), sometimes they integrate a variety of governmental activities, and sometimes they operate completely independent of government departments (i.e., a local nonprofit organization). In any case, sustainable communities initiatives have emerged and exist in cities all around the country. What these initiatives do—how they define and attempt to achieve their objectives is what this research is all about. As stated earlier, it is quite premature to try to provide any systematic assessment of whether these initiatives have made their cities more sustainable. Undoubtedly, that will be an appropriate issue to study in the years to come. But, perhaps it is not premature to assess whether there are cities that do seem to be taking urban sustainability seriously, and that is what this research attempts to do here.

Much of the analysis that follows is based on information about a number of cities in the United States. The central focus is on a handful of cities that seem to have established themselves as true innovators in their efforts to become sustainable. These cities—including Seattle, San Francisco, Chattanooga, Scottsdale, Portland, Santa Monica, Cleveland, Boulder, Austin, New Haven, Jacksonville and Tampa, San Jose, Boston and Cambridge, and Santa Barbara—are perhaps the most frequently cited examples of places that have taken significant sustainability

Table 1.1
List of cities with sustainability initiatives

City	Name of sustainability effort
Chattanooga, TN	Sustainable Chattanooga
Jacksonville, FL	Jacksonville Indicators Project, Jacksonville Community Council
Orlando, FL	Sustainable Communities
Tampa, FL	The Tampa/Hillsborough Sustainable Communities Demonstration Project
Seattle, WA	Sustainable Seattle/The Comprehensive Plan
Olympia, WA	Sustainable City Indicators/Sustainable Community Roundtable
Portland, OR	The Comprehensive Plan
Milwaukee, WI	Campaign for Sustainable Milwaukee
Santa Monica, CA	Santa Monica Sustainable City Program
San Francisco, CA	The Sustainability Plan
San Jose, CA	Sustainable City Programs (Sustainable City Major Strategy, part of San Jose 2020)
Santa Barbara, CA	The South Coast Community Indicators Project
Austin, TX	Sustainable Communities Initiative and Sustainability Indicators Project of Hays, Travis, and Williamson Counties
Indianapolis, IN	IndyEcology
Boulder, CO	The Sustainability Program
Cambridge, MA	Sustainable Cambridge, Cambridge Civic Forum
Boston, MA	Sustainable Boston Initiative
Brookline, MA	Comprehensive Plan
Scottsdale, AZ	Scottsdale Seeks Sustainability
Tucson, AZ	The Livable Tucson Vision Program
Phoenix, AZ	Comprehensive Plan, Environmental Element
Brownsville, TX	Eco-Industrial Park
Cleveland, OH	Sustainable Cleveland Partnership, EcoCity Cleveland
Lansing/East Lansing, MI	Sustainable Lansing
Ithaca, NY	EcoVillage at Ithaca
Burlington, VT	No apparent name
New Haven, CT	Vision for a Greater New Haven
Annapolis, MD	Alliance for Sustainable Community
Oklahoma City, OK	Possibilities: Neighbors in Action
Grantsville, UT	Grantsville General Plan for Sustainable Community
Stuart (Martin County), FL	Sustainable Community

initiatives. But the analysis is not limited to just these cities. In some systematic way, this analysis attempts to address the question of whether these cities really are making greater strides than other cities that haven't necessarily received the same level of notoriety. Are they really taking sustainability more seriously? Instead of assuming that they are, based on anecdotal information, this analysis wishes to treat this as a research hypothesis. Ultimately, the analysis turns its attention toward trying to understand what some cities do that would seem to place them on the path toward achieving significant sustainability goals, and why they have progressed as far as they have. Why do some cities take sustainability more seriously than others?

Trying to assess how seriously cities seem to be taking the pursuit of sustainability is no small task. The conceptual literature on sustainability, as discussed earlier, is not altogether useful as a foundation for making judgments about what cities are doing. But the local context of sustainability adds another layer of complications. Not only do different cities face different sustainability challenges, they also possess very different capacities for dealing with these challenges. All cities in the United States tend to have experienced some amount of urban sprawl over the last ten to twenty years, but sprawl as a particular characteristic of urban growth affects some cities far more than others. Moreover, whether, in what way, or to what degree cities are able to manage the growth of their metropolitan areas is likely determined by a combination of local resources and the type of urban regime that governs. The metropolitan context of cities and the nature of their urban regimes set the stage for understanding cities' needs and capacities for taking sustainability seriously.

The Metropolitan Context of Sustainable Cities

Although cities represent distinctly separate units of government in the United States, they do not exist in a vacuum. Cities exist in a larger geographic context usually described as a *metropolitan area*. In fact, the larger context in which cities exist can be quite varied from city to city. Some cities, particularly older northeastern cities such as Boston and New Haven, serve as the core of relatively high-density economic and social activity for a much larger area of smaller cities and suburban towns. Other cities, particularly newer cities including many in California, are more similar to their surrounding communities in terms of population density. Some cities, such as Austin, Texas, represent the dominant population center in the larger area. And still other cities, such as Jacksonville, Florida, are the metropolitan area. Due to the consolidation of the City of Jacksonville with Duval County, the city contains virtually all of the metropolitan area, forming one geographically huge and sprawling place.

The point is that the larger geographic context in which cities exist can be quite varied. Indeed, cities vary considerably in the extent to which there is an active, metropolitan-wide sustainability effort to augment or reinforce cities' initiatives. Some cities that are trying to pursue sustainability initiatives are able to do so in the context of parallel efforts by larger governing or planning entities. For example, the sustainability efforts in Chattanooga are complemented by specific programs in Hamilton County (such as the Chattanooga–Hamilton County Air Pollution Control Bureau, and the Hamilton County "Indicators of Community-Well Being" project), the Tennessee Valley Authority, and various projects of the metropolitan-wide Chamber of Commerce. While Austin, Texas, operates its own sustainability initiative, the sustainable indicators initiative of Travis, Williamson, and Hays Counties establishes the larger context there. And for cities in California, such as Santa Monica, San Francisco, and San Jose, and in Washington State, such as Seattle and Olympia, statewide comprehensive growth management mandates require that cities' efforts be part of metropolitan-wide planning projects.

The reason why the metropolitan context for cities engaged in the pursuit of sustainability promises to play an important role is largely due to the fact the larger geographic area in which cities exist manifests the immediate externalities that the city government must contend with. While cities may well have the legal authority to address issues and problems within the limits of their geographic boundaries, they tend to have much less ability to affect externalities imposed on them from outside. For example, cities sometimes find themselves embedded in a metropolitan area dominated by sprawling residential and commercial development that affects the quality of their environment. Such sprawl may produce undesirable consequences, and the city may feel powerless to affect it. How the character of cities' metropolitan areas affects, influences, and constrains cities' efforts to become more sustainable is examined throughout this analysis.

Urban Regimes and the Seriousness of Sustainable Cities

Scholars of urban politics have long recognized the importance of what they call "urban regimes" in determining and defining the policies and programs pursued by cities. Perhaps starting with Clarence Stone's (1989) path breaking study of Atlanta, the idea has gained acceptance that local policies are the product of particular types of *informal* arrangements among political actors or stakeholders in the city, rather than simply the policies pursued by elected officials operating alone. To Stone, an urban regime is defined by the nature of the relationships, the informal partnership, between city hall and the "downtown business elite." As he states it:

> An urban regime may . . . be defined as the informal arrangements by which public bodies and private interests function together in order to be able to make and carry out governing decisions. . . . They have to do with managing conflict and making adaptive responses to social change. The informal arrangements through which governing decisions are made differ from community to community . . . (Stone 1989, 6)

Stone outlines in great detail the various kinds of informal arrangements and underlying processes that operated in Atlanta. Stone and others have applied the concept of urban regimes to try to understand the foundations for building coalitions in support of, or opposition to, a variety of policies and programs related to sustainability, particularly growth management or growth control efforts (Fainstein and Fainstein 1986; Leo 1998; Stone 1993). Stone developed a typology of regimes based on the nature of what kind of governing responsibility the city has elected to undertake. His four regime types include the "maintenance regime," which is primarily focused on the basics of city government, routine service delivery. Some cities concentrate almost exclusively on such functions, opting not to try to do more. A second, more ambitious regime type, the "development regime," seeks to marshal local resources and coordinate local institutional elites for the purpose of pursuing traditional economic growth and development. The third regime type, what he calls the "middle class progressive" regime, focuses on such measures as environmental protection, historic preservation, affordable housing, the quality of urban design, affirmative action, and linkage funds for various social purposes (Stone 1993, 19). And the fourth regime type, the "lower class opportunity expansion" regime, seeks to achieve substantial equity and resource distribution at levels not found in very many U.S. cities.

The important distinguishing character of the different regime types, besides the differences in their ultimate goals, is the type and amount of resources necessary to achieve these goals. In effect, the four regime types are arrayed from the easiest to the hardest to achieve. In particular, moving from the development regime to the middle-class progressive regime turns out to be an enormous political and managerial challenge. Since the pursuit of sustainability would require cities to move from being maintenance or development regimes to middle-class progressive or lower-class opportunity expansion regimes, depending on whose definition of sustainability one uses, the political challenge for advocates of sustainability is formidable, to say the least.

Although a comprehensive study of the relationship of urban regimes and the pursuit of sustainable cities far exceeds the scope of this analysis, there appear to be some aspects of the relationship between sustainability initiatives and urban regimes that are worth exploring. In many cities with some form of sustainable cities program, the initiative to pursue sustainability came from informal arrangements of local elites, nonprofit organizations, business groups, including chambers of commerce, elected and appointed officials, and sometimes neighborhood associations. It is not at all clear that there is any sort of "model" or common form of such arrangements that maximizes the chances that the city will be able to take sustainability seriously. Indeed, while many local advocates of sustainable communities pursue particular strategies under the assumption that they will be able to alter the urban regime, such strategies do not seem to produce a strong record of success. The ways in which the urban regimes of various cities have been altered or affected in the pursuit of sustainability will be addressed throughout this book, particularly in chapters 5 and 7.

The Plan for the Book

In the next chapter, this analysis turns its attention to the "measurement" of sustainable cities. Measurement refers to the kinds of information that can be taken as evidence of a city's seriousness about trying to achieve greater sustainability. Based largely on the policies, programs, and activities of cities, including sustainable indicators initiatives, this discussion focuses on the broad range of issues that cities have addressed in their efforts to become more sustainable. The focus here is on what kinds of issues are necessary to be included in sustainable cities initiatives if they are serious. Clearly, deep concern for the biophysical environment and ecology of the city is a core element in, or dimension of, taking sustainability seriously. Cities can be assessed on this dimension in terms of the extent to which they possess such deep concern. But the second element, whether and the extent to which there is also deep concern for livability and quality-of-life issues, and a connection between conceptions of the biophysical and quality-of-life, provides an additional dimension on which cities can be assessed. A number of the cities that seem to take seriously the ecological and biophysical dimension of sustainability also take seriously the connection of ecological issues to livability issues. But there are some cities where this connection is notably less direct or clear-cut. From a political perspective, there is little doubt that sustainable cities initiatives constitute a vehicle for progressive advocates to press their agendas for social and human services reforms. This advocacy finds a friendly home in the nexus between the biophysical environment and more general quality-of-life issues. Yet, in many respects, this nexus is a very hard pill for local policymakers in some cities to swallow.

The purpose of subsequent chapters is to highlight specific dimensions of cities sustainability initiatives, starting with the environmental and energy dimension in chapter 3, followed by the economic growth and development, or "smart growth," dimension in chapter 4, the participatory and community-building dimension in chapter 5, and the environmental and social justice dimension in chapter 6. Each of these chapters discusses the general concept underlying its contribution to sustainable cities, and highlights the activities, programs, and policies that are part of specific cities' sustainability initiatives.

Subsequently, the analysis turns its attention to examining the activities and initiatives in a number of cities themselves, providing a somewhat more comprehensive picture of what cities are doing. Chapter 7 focuses on eight cities—Chattanooga, Jacksonville, Austin, San Francisco, Santa Monica, Portland, Seattle, and Boulder—that clearly seem to have crossed the threshold in the sense that they comfortably and explicitly confront issues of sustainability, often including issues of community building and environmental and social justice. Chapter 8, by contrast, largely focuses on cities that have some impressive sustainability activities or initiatives but that do not seem to have quite so comfortably or comprehensively crossed the threshold. It also contains an assessment of patterns across twenty-four of the cities that are listed in table 1.1.

As a result, this contrast illuminates a fundamental issue concerning whether any city is, or all cities are, capable of taking sustainability seriously. Although this issue has not previously been highlighted in the literature on sustainable communities, one might observe some elements of political ideology and ideological bias implicit in many definitions of sustainable communities. If a city must be a communitarian kind of place, and must incorporate true concern for environmental justice, before it can be said to be seriously working toward achieving sustainability, what does this imply for the future of sustainable cities? Does this suggest that cities whose dominant political ideologies are not progressive or liberal cannot take the pursuit of sustainability seriously? Additionally, what kinds of national and state policies would seem to be conducive to taking sustainable cities serious? These are the central questions raised in chapter 8.

2

Measuring the Seriousness of Sustainable Cities

One purpose of the analysis in this book is to assess whether and in what ways some cities seem to take issues of sustainability more seriously than others. A part of any assessment effort must include the development of the standards and criteria by which cities may be judged. Given the conceptual underpinnings discussed in chapter 1, is it possible to begin evolving these standards and criteria? Keeping in mind that the task at hand is not to assess whether cities have become more sustainable, for that is a task that is both premature and perhaps presumptuous to undertake here, and given current understandings of what sustainability conceptually means in an urban setting, is it nonetheless possible to articulate a broad range of criteria and standards that help to define whether any given city is taking sustainability seriously? This chapter begins the process of investigating the operationalization of the concept of "taking sustainability seriously." As it turns out, this is not an easy task to accomplish. The concept of "taking sustainability seriously," like the concept of sustainability itself, is complex, multidimensional, and perhaps even situational. Yet the question lingers whether there are some basic elements to any serious local sustainability effort.

The process of establishing criteria and standards cannot avoid being fairly complicated. From the outset, as discussed in chapter 1, there are numerous conceptual variations of sustainability, each carrying its own prescription for more concrete definitions. Any close reading of the conceptual literature leads to the conclusion that sustainability itself is a multidimensional concept, and any effort to measure how seriously a city seems to take sustainability would necessarily also have to account for its many dimensions. Additionally, applying the same standards across cities must be done carefully because cities differ in the specifics of the kinds of needs they have for programs, policies, and actions. For example, because of severe air pollution problems, air pollution abatement efforts represent a key part of the sustainability initiatives in Chattanooga. Not every city has the same level of need to address air pollution, per se, so on its face it would seem unreasonable to judge all cities by the same standard. If these impediments were not enough, the challenge here is not, strictly speaking, to operationalize sustainability, per se, but to focus on how seriously cities take their pursuit of sustainability. The task is not to develop measures of how sustainable cities are, but rather how seriously they seem to take the quest for sustainability. However, this does not mean that standards and criteria for assessing the seriousness of cities' sustainability initiatives are impossible to establish. Indeed, this chapter discusses a variety of ways that this task can be approached.

In practice, the key distinguishing feature among cities—the characteristic that differentiates more serious from less serious cities—is whether issues of sustainability can be said to be clearly and unambiguously on the public agenda. In general, agenda setting is a broad and multidimensional issue involving a sometimes lengthy and difficult to observe process. Although the underlying agenda-setting processes is discussed throughout this book, suffice it to say that this does not represent a comprehensive effort to understand cities' agenda-setting processes, per se. Rather, this analysis focuses on one specific element of agenda setting, the search for tangible evidence of how important sustainability is to cities especially as a matter of public policy. Consequently, a major focus here is on whether cities have developed and established some type of officially recognized sustainability plan. By now, virtually all cities in the United States have developed specific environmental programs, particularly household recycling programs, brownfields redevelopment, water management and conservation, and perhaps hazardous materials management as well. Arguably, such programs can be considered efforts to reduce the ecological footprint of the city. But some cities have gone well beyond these narrowly targeted and often piecemeal programs to try to be more comprehensive and inclusive. Sustainability plans typically serve as the vehicles for becoming more comprehensive. So much of the discussion that follows examines the content of cities' sustainability plans.

Why a Single Index?

Perhaps in the best of all possible worlds, analysis of the seriousness of cities' sustainability efforts could be facilitated through the development of a single index, or measure, that captures in some appropriate way all of the various dimensions of sustainability. Perhaps much like the "Green Metro Index" developed by the World Resources Institute (1993), the Green Index of environmental policies and programs in U.S. states (Hall and Kerr 1991), or the more recent efforts by Dan Esty and colleagues (2001) to develop an "Environmental Sustainability Index" for the nations of the world, a single "taking sustainability seriously index" could be developed. An effort is made here to develop a very rudimentary index, one that focuses on the elements of what it means for a city to take sustainability seriously, based on the range of policies, programs, and other actions that have been enacted and/or implemented in cities. The emphasis here is not on the extent to which cities have actually achieved particular environmental or livability results. Among the various efforts to measure sustainability, there is some precedent for this approach. Indeed, while the Green Metro Index is largely based on data measuring the actual quality of the environment and consumer behaviors in metropolitan areas of the United States, Esty's work contains a significant dose of institutional capacity to affect environmental quality, and numerous indicators of actions taken by national governments that might improve or protect the quality of the environment in the future.

Esty's "Environmental Sustainability Index" for nations was computed based on sixty-seven variables, composing twenty-two "core indicators" or dimensions of sustainability. While most of these variables represented attempts to quantify the quality of the environment (for example, nitrogen oxide emissions per populated land area, pesticide use per hectare of agricultural crop land, etc.), many other variables represented attempts to measure the capacity of the nation to deal with sustainability issues. For example, among the sixty-seven variables are measures of the "stringency and consistency of environmental regulations," the "degree to which environmental regulations promote innovation," and the "number of sectoral environmental impact assessment guidelines" in use (Esty 2001, Annex 4 and 6). All of these variables are measured with seven-point attitudinal scales from surveys. None of these latter variables, or many others contained in the study, directly measures the quality of the environment, per se. Instead, they measure elements of how seriously the nation seems to take the pursuit of sustainability.

The essential problem with developing a single index of "taking sustainability seriously" is that neither the conceptual nor the empirical work on sustainability has been able to link, with any confidence, specific actions, policies, or programs to a clean or an improving environmental quality. In other words, we simply do not know how much any specific program or municipal action might be said to contribute to becoming more sustainable. When a city adopts a residential recycling program, and an estimated 40% of residential waste is in fact recycled, how much does this contribute to sustainability? This is a difficult question to answer, although it is not impossible to imagine a way to produce an estimate. Indeed, that is precisely what Rees (1992, 1996), Rees and Wackernagel (1994), and Wackernagel (1998) have tried to do in the development of their "ecological footprint" measure, as discussed later in this chapter. Essentially, they convert the environmental impact into a single measure, an amount of land required to support the lifestyles and consumption of the city's residents. Presumably, a city that recycles 40 tons of household solid waste would have a smaller footprint than one that recycles 30 tons of waste, all other things being equal.

Yet this approach cannot readily extend to municipal policies or actions that do not have an immediate measurable impact on the environment. A city that is well along the planning process for a major eco-industrial park might be said to take sustainability more seriously than a city that has no such plans. Yet until the park is operational, it produces no particular improvement in the environment. A city that has integrated sustainability concerns into its entire municipal planning process might be said to take sustainability more seriously than a city whose programs are created piecemeal in uncoordinated fashion. Until the municipal actions produce measurable results in the environmental, there is no way to know how much each action can be said to contribute to taking sustainability seriously. At best, one must make assumptions that such city policies and programs will produce desirable environmental outcomes, but ascertaining how much and over what period of time would simply involve guesswork. At some point, perhaps ten or twenty years down the line, researchers will be able to estimate how much each of a large number of actions might be said to contribute to actual environmental improvements or to diminished environmental degradation. Until then, the creation of a single index is inevitably hampered with problems. It is nonetheless important to understand all the elements that must necessarily go into any assessment of how seriously cities seem to take sustainability.

The Focus of Analysis: Sustainable Cities Initiatives

Before an effort is made to examine what cities are doing to try to become more sustainable, it is necessary to clarify with a little more precision what the focus of this analysis is. As discussed in chapter 1, the central focus here is on the city. Although much of the discussion to come provides examples of specific kinds of programs, policies, and activities that are consistent with the pursuit of sustainability, the principal focus of this analysis is on what might be called "the sustainability initiative." A sustainable cities initiative is operationally defined here as any set of activities, programs, policies, or other efforts whose purpose is explicitly to contribute to becoming more sustainable. Such initiatives could be operated by a city government agency. They could be threaded through the operation of all of city government. They might be initiated and/or operated independently of city government by a nonprofit organization or foundation, although, as later discussion addresses, there is some question as to whether a sustainability initiative that does not include city government agencies can really be said to be a serious one.

Sustainability initiatives come in many shapes and sizes. They have many different features, characteristics, and elements, and many of these features are tailored to the specific environmental and political realities of the place in which they are developed. The first order of business, then, is to begin the process of explicating what some of these features and characteristics are. Instead of considering what cities might be able to do in the abstract or in the best of all possible worlds, this discussion is based on an understanding of what cities are actually doing. Once the basic characteristics of sustainability initiatives are described, an effort can be made to examine the connection between these characteristics and the degree to which cities seem to take sustainability seriously.

Sustainability Plans

Perhaps the single most important element in assessing the seriousness of a city's efforts toward achieving sustainability is the presence of a sustainability plan. Principally because most of the concepts of sustainability perforce require more holistic ecological views, the presence of a sustainability plan signals the city's willingness to address numerous issues in a systematic way. Frequently, such plans will outline a variety of different issues or problems that need to be addressed, and identify the administrative and organizational resources available to address them. For example, Chattanooga's sustainability plan focuses on clean water issues, land and forest conservation, energy and transportation, recycling and materials recovery, and industrial and economic development, particularly the use of eco-industrial parks (Sustainable Chattanooga 1995). Sustainability plans vary in the extent to which they prescribe integrated, as opposed to insular, approaches to address the problems. The sustainability plans that appear most fully developed are those that take the form of "strategic plans" that incorporate "indicators of sustainability."

In some instances, city governments may incorporate sustainability objectives into broad comprehensive strategic plans. Seattle, Washington, and Portland, Oregon, provide clear cases in point. Sustainability goals of the sort that were developed by the nonprofit Sustainable Seattle organization are threaded throughout the government's "Comprehensive Plan," which is designed to provide a detailed twenty-year strategic vision for the city. This plan contains elements on the environment, land use, economic development, transportation, and many other areas. In Portland, Oregon, sustainability goals are contained within the city's 1999 "Comprehensive Plan," which integrates provisions for transportation, energy, economic development, and the environment (City of Portland 1999). Seattle's Comprehensive Plan is even subtitled "Toward a Sustainable Seattle," and contains sustainability goals throughout (City of Seattle 2000).

Perhaps one of the more important distinguishing characteristics among sustainability plans or integrated strategic plans is the inclusion of "indicators of sustainability." Numerous efforts have been made to develop plans that include specific indicators, or measures, of progress toward achieving sustainability, many of which have been developed in, by, and for specific communities. According to Beatley and Manning (1997, 203), as of 1997 there were at least forty communities in the U.S. that had developed explicit sustainability indicators, and judging from the numbers of case studies of communities, there are many more than that. As of early 2001, the *Global Cities Online Project*, for example, lists some sixty-four case examples of indicators projects (Global Cities Online 1998). Not all of those projects are for cities, but there are many cities among them. The primary interest here in these sustainability indicators is not to use them to assess whether cities have achieved their goals and become more sustainable. Since so little time has elapsed since cities began their pursuit of sustainability, such an effort would be premature. Rather it is to examine them to see what they consider the important standards and criteria for working toward sustainability, and to see whether they provide an appropriate outline of what kinds of issues cities need to address if they wish to take sustainability seriously.

The purpose of this discussion is to examine the broad range of issues that have been raised as appropriate measures of progress toward achieving sustainability, and to discuss whether it is equally appropriate to use these as measures of how seriously cities take achievement of sustainability. This discussion shows that the ways that cities define sustainability is quite varied, and we eventually discover that indicators of sustainability developed in some places often go far beyond those that directly relate to the physical environment of cities. An outline of the areas or elements covered by sustainability indicators in cities is presented in table 2.1. Frequently, sustainability indicators "projects" include measures of a variety of social and political characteristics, such as whether there is widespread public participation in city affairs, social equity, economic equality, or concern for diversity, among many others. Looking at the broad range of characteristics thought to be important elements of sustainability in cities, this chapter begins the process of sorting out those that would seem to reflect greater seriousness in their efforts to achieve sustainable goals. Subsequent chapters will focus on some of these specific indicators that transcend the physical environment—those related to the pursuit of communitarian goals, environmental justice, and constrained economic development.

Table 2.1
Elements of serious sustainability indicators in cities

Some key Sustainability Indicators elements	In what way does this element reflect the seriousness of the sustainability effort?
Is there an Indicators Project?	An Indicators Project reflects a more serious effort
Is the Indicators Project part of a Strategic Plan or a Comprehensive Plan?	If the Indicators Project is part of a larger Strategic or Comprehensive Plan, this represents a more serious effort
Is there a citywide Comprehensive Plan that integrates or incorporates environmental issues?	If a Comprehensive Plan incorporates sustainability goals, this represents a more serious effort
Does the Indicators Project or the Comprehensive or Strategic Plan include indicators of:	
Environment, ecosystem health, ecological footprints?	Presence of indicators of environmental integrity are essential to a serious effort
Energy efficiency/consumption?	Presence reflects efforts to monitor usage and reduce pollution from fossil fuel burning energy sources
Local economic development?	Economic development indicators tend to reflect concern for sustainable development
Quality of life?	Use of quality of life indicators suggest a more serious effort, although there is disagreement about this
Environmental equity/social justice?	Equity indicators generally reflect a more serious effort, although their absence may simply reflect the difficulty of measurement and lack of data in this area
Governance?	Indicators of governance reflect a more serious effort, although absence of them may simply reflect the difficulty of measurement and lack of data in this area
Are there action plans associated with the Indicators Project or with the Strategic Plan?	Presence of action plans represent commitment to activities that must be accomplished, and reflect a more serious effort
Is the Sustainability Initiative operated by a single high-level, centralized, administrative agency or department a of city government rather than a nonprofit organization, a low-level, highly specialized functional city agency, or dispersed across numerous functional agencies?	A single agency implies clear delineation of responsibility and accountability; complete reliance on the nonprofit sector or specialized city agencies makes implementation more difficult; dispersed responsibility implies less coordination and accountability for sustainability results
Is the Sustainability Plan or Indicators Project connected to an area-wide (county, metropolitan, regional) plan?	Connecting the city's plan to an area-wide plan facilitates coordination and provides the opportunity to reduce the city's externalities

The presence of an indicators project in a city, by itself, does not imply that sustainability is being taken truly seriously. Indeed, as is discussed later, there are numerous examples where cities have well-developed indicators projects, but these projects do not include, or are not combined with, any explicit efforts (policies or programs) to improve sustainability. In other words, the indicators may be used to measure success or failure at becoming more sustainable, but may not carry with them specific plans of action concerning how to become more sustainable. This is particularly noticeable when a city's indicators project reports that things have become worse in some area. The next obvious question is "what do we do about that?" In the absence of a plan of action, it is often difficult for a city to know what to do in response to an indicator that moves in the wrong direction over time. When a city with a series of indicators on air emissions, for example, finds that air pollution is getting worse, the city must look beyond the indicators project to formulate possible remedies. Yet even within the context of indicators projects themselves, there is a wide array of characteristics that cities elect to include as measures of sustainability.

The processes used in specific cities to develop sustainability plans are varied as well. Each city puts its unique imprint on the process according to where the locus of the planning takes place, who is involved, and what issues are or are not placed on the agenda. In some cities, sustainability plans emerge from the nonprofit sector. In other cities, they come out of an active city agency, such as a planning department. Sometimes government officials are intimately involved, and sometimes only tangentially so. In some places, the business community is at the heart of the effort, and in others it is held, or elects to stay, at arm's length. Details about the processes that have been associated with specific cities' sustainability plans are discussed later. Suffice it to say that the eventual agenda of any sustainability plan is in large part a product of the values and agendas that participants bring to the planning process.

Indicators of Sustainability and the Quality of Life

Over the last ten years, many efforts have been made to create rather pragmatic indicators of sustainability and the quality of life that can be applied in various local, regional, or national settings. Cities such as

Seattle, Santa Monica, Austin, Santa Barbara, Boston, Jacksonville, and others, have engaged in the process of initiating "indicators projects" to provide guidance in their respective efforts to become more sustainable. These indicators represent collections of measurable characteristics that can be tracked over time. These indicators have been developed for the purpose of providing reasonably clear measures and benchmarks for assessing whether progress is being made toward becoming more sustainable. For example, a city may wish to make progress toward reducing its air pollution, and select indicators based on the particles per million in the air of specific pollutants, such as sulfur dioxide and carbon monoxide. Once selected, the city can attempt to measure the presence of the pollutants at regular points in time. Alternatively, indicators could be established that focus on point sources of pollution, such as how many new cars registered in the city use alternative fuels (electric vehicles, low-emission vehicles, etc.). As a result of the indicators, the city can measure and document whether, and to what extent, it is making progress toward its sustainability goal of reducing air pollution. Presumably, each selected indicator has a "good direction" or a "bad direction," where movement of an indicator over time in a good direction suggests progress toward greater sustainability. As Beatley and Manning (1997, 205) suggest, ". . . benchmarks or targets, as natural extensions of indicators, provide a tangible and specific goal for sustainable places and an ability to know when a community is being successful and when it is falling short."

Not all sustainable indicators projects are created equal. In some cities, indicators efforts are developed by local nonprofit organizations alone. The indicators project in Olympia, Washington, part of its "Millenium Project" [sic], for example, was developed by its Sustainable Communities Roundtable, a local nonprofit organization. Sometimes they are developed in conjunction with a city agency, such as a planning department. Sometimes a city agency establishes the indicators project and develops the indicators, as has been the case in Santa Monica, California. Sometimes indicators projects are the product of parallel independent efforts of citizen groups and city agencies, as happened in Seattle. And sometimes a city will farm out the process of developing indicators to a private company. In Tampa, Florida, for example, the city hired a consulting company from Portland, Oregon, to apply its computerized packaged sustainability framework to initiate a plan for Tampa. There is also a great deal of variation in the goals of the indicators projects themselves. Some see indicators projects as producing sets of measurable results that can be used essentially as internal management tools—as part of an overall effort to implement a sustainability plan. Indicators projects can also be viewed as encompassing a process whose goal is to engage the city's general public, to tap directly into the values and attitudes of the city's residents to generate a grassroots "visioning" statement of what makes the city sustainable or livable.

Indicators projects essentially scour existing sources of information available for their cities, and consider which pieces of information potentially provide some insight into progress toward sustainability. Rarely does an indicators project seek to develop new measures of characteristics they think are important. In other words, an indicators project would not likely define a measure that required some new data collection initiative not already undertaken, although there are few exceptions. Projects typically rely on information available from various other federal, state, or local governmental agencies. For example, population and housing characteristics often are taken from the U.S. Bureau of the Census, employment and unemployment information from the U.S. Bureau of Labor Statistics, and environmental quality data from the regional office of the U.S. Environmental Protection Agency or a state or regional environmental agency (an air quality district, for example). Sometimes local government agencies contribute information. For example, information about the city's household participation in recycling, or the amount of solid waste that is recycled would typically come from a public works or other city agency that runs its recycling program. Thus, projects often adopt as indicators characteristics that are relatively easy to measure, i.e., where existing data already exist.

Cities that have developed indicators of sustainability usually incorporate a broad range of factors into their measures. In San Francisco, for example, as outlined in box 2.1, sustainability indicators include measures of air quality, biodiversity, energy, climate change, and ozone depletion, food and agriculture, hazardous materials, human health, parks and open spaces, economic development, environmental justice, education, and much more. In Seattle, as depicted in box 2.2, indicators span perhaps an even wider range of considerations. Other cities, such

Box 2.1
Indicators of Sustainability from San Francisco

Environment
Number of existing buildings that join the Building Air Quality Alliance
    Program (or similar voluntary programs)
Number of people going to clinics for respiratory problems
Percentage of new cars registered in San Francisco that are alternatively
    fueled (e.g., California Air Resources Board-certified, low emission
    vehicles, ultra-low emission vehicles, or electric vehicles)

Biodiversity
Number of volunteer hours dedicated towards managing, monitoring, and
    conserving San Francisco's biodiversity
Number of square feet of the worst invasive species removed from natural
    areas
Number of surviving indigenous native plant species planted in developed
    parks, private landscapes, and natural areas
Abundance and species diversity of birds, as indicated by the Golden Gate
    Audubon Society's Christmas bird counts

Energy, climate change, and ozone depletion
Ratio of renewable to nonrenewable energy consumption
Energy cost per tax dollar

Food and agriculture
Number of public agricultural gardens
Quantity of food and agricultural residuals recycled
Number of school, vocational, and community education and training pro-
    grams about sustainable agriculture and nutrition

Hazardous materials
Difference between motor oil purchased in the city and the amount that
    is properly recycled or disposed
Equitable distribution of the hazardous material/waste exposure load
    throughout the city
Number of contaminated sites within city borders
Public awareness of hazardous materials/waste issues (especially proper use
    and disposal and knowledge of alternatives) as measured by annual
    survey

Human health
New cases of asthma
Number of people attending organized wellness classes
Participation in organized youth programs at city recreation centers

Parks, open spaces, and streetscapes
Percentage of the population with a recreational facility and a natural
    setting within a ten-minute walk
Number of neighborhood green street corridors created annually
Number of volunteer hours spent annually on maintenance of open space
Annual municipal expenditures on parks, open space, and streetscapes

Box 2.1
(continued)

Solid waste
Tons of waste landfilled annually
Recycling rate as a percentage of material generated
Percentage of residents, businesses, and institutions that participate in recycling programs

Transportation
Auto registration
Parking-spot inventory
Muni ridership
Muni route running time on key routes

Water and wastewater
Per capita water consumption measured by the San Francisco Water Department
Mass of pollutants in wastewater
Mass and frequency of combined sewer overflows
Recycled water use
Acres of habitat restored

Economy and economic development
Number of San Francisco enterprises adopting ISO 14000 standards
Number of San Francisco neighborhoods with unemployment rates higher than the government-defined "full employment" rate
Difference between the highest neighborhood unemployment rate and the full employment rate
Number of San Francisco manufacturers using recovered secondary materials as raw material
Percentage of people employed in San Francisco who live in San Francisco

Environmental justice
Mean income level of people in historically disadvantaged communities
Proportion of environmental pollution sources in historically disadvantaged communities with respect to San Francisco's other communities
Participation of historically disadvantaged communities as a whole and their indigenous self-selected representatives in decision-making processes

Municipal expenditures
Number of items of legislation adopted by the Board of Supervisors that advance sustainability goals
Number of service providers and companies on the Green Vendors list
Percentage of budget allocated utilizing sustainability criteria
Percentage of budget that is devoted to facility maintenance

Public information and education
Number of schools that integrate and progressively update environmental education in their curricula

Box 2.1
(continued)

Conservation and waste reduction as measured by volume of garbage pro-
    duced per capita and units of electricity used per capita
Number of volunteers working on environmental projects as measured
    through the largest volunteer clearinghouse that refers or mobilizes
    people to do community service
Risk management (activities of high environmental risk)
Number of businesses that train employees in the Neighborhood
    Emergency
Response teams program
Number of seismically upgraded buildings
Number of hazardous materials incidents

Source: Sustainability Plan for San Francisco—October 1996. Found at
<http://www.ci.sf.ca.us/environment/sustain/Indicators.htm>.

as Jacksonville, which develop their indicators as part of an explicit
"quality of life" initiative include a wide array of measures, including
the quality of housing, the crime rate, and many other characteristics of
the lives residents experience.

Obviously, the establishment and use of indicators of sustainability
serve the primary purpose of providing targeted measures and benchmarks for cities to judge their own progress toward achieving sustainability. But the development of sustainability indicators by communities
may also serve other purposes. In the broadest sense, such community
projects not only serve as articulations of what cities think is important,
they also may imply some planning process that can be used to coordinate city departments that are typically unconnected, and to involve
citizens and residents in that process. According to Elizabeth Kline, as
reported by Jill Zachary of the Community Environmental Council, Inc.,
in Santa Barbara, the development of sustainability indicators has the
effect of:

· Enabling a community to identify what it values and prioritize those
values

· Holding individuals and a larger group accountable for achieving the
results they want

• Democracy building; through collaboration people engage in a community-building process.
• Allowing people to measure what is important and make decisions based on those results. They measure whether we are achieving what we want and whether the outcomes are improving our lives. (Zachary 1995, 7)

Of course, nothing guarantees that the development of indicators will produce these results, and whether, or the extent to which, the development of sustainability indicators in a particular community actually performs any of these functions may well depend on how seriously the community takes the process and its goals. In its most serious form, sustainable indicators initiatives resemble full-scale strategic planning processes, where goals are set, indicators and benchmarks are selected, actions necessary to meet the goals are specified, and the entire apparatus of city government gets behind the enterprise.

While such indicators are almost always developed for the purpose of tracking progress toward sustainability, the very fact that a city has elected to try to measure progress on reducing air pollution, for example, tells us much about that city. It tells us that the city is concerned with its air quality, perhaps with the amount of pollution that is being emitted by human activity in that city, and perhaps the effects on the city of human activity outside the city. It also tells us that the city has started to take seriously the issue, and has taken systematic steps to begin understanding the consequences of human activity in the city. To the extent that these efforts are coupled with explicit initiatives and actions to reduce air pollution, it tells us that there is a serious effort to reduce the ecological footprint of the city, at least with respect to air pollution.

Just as air pollution can be the target of sustainability indicators, so too can many other environmental, social, and political characteristics. Each indicator used by a city is itself a reflection of what it considers important to address. Taken as a whole, a collection of indicators provides a sense of the city's overall view of what it considers sustainability to be. Should a city focus just on air pollution, or should it look at a variety of types of environmental pollution? In general, one might infer that a city that decides to develop indicators for multiple types of pollutants takes sustainability more seriously than one that focuses only on air pollution. However, as is discussed later, such an inference may not

Box 2.2
Sustainable Community Indicators from Sustainable Seattle

Environment
Wild salmon returning to spawn in King County streams
Wetlands health as measured by water quality, water level fluctuation, and
  amphibian health
Bidiversity as measured by amphibian and plant diversity in King County
  wetlands
Soil erosion as measured by turbidity levels in King County waterways
Percentage of Seattle streets meeting pedestrian friendly criteria
Impervious surface are in the City of Seattle
Air quality as measured by the EPA Pollutant Standards Index
Open space as measured by acres of accessible open space

Population and resources
Population growth rate
Residential water consumption per capita
Solid waste generated and recycled per capita
Pollution prevention and renewable resource use as measured by the EPA's
  Toxic
Release Inventory and use of recycled paper products
Farm acreage in King County
Vehicle miles traveled and fuel consumption per capita
Renewable and nonrenewable energy use per capita

Economy
Percentage of jobs concentrated in top ten employers
Real unemployment
Distribution of personal income
Health care expenditures per capita
Hours of work at King County average wage required to meet basic living
  needs
Housing affordability
Percentage of children living in poverty
Emergency room use for nonemergency health care purposes
Community capital as measured by total and per capita deposits in local
  banks

Youth and education
Adult literacy
High school graduation rate
Ethnic diversity of teachers in public schools
Arts instruction
Volunteer involvement in schools
Juvenile crime rate
Number of youth involved in community service

Health and community
Equity in the justice system as measured by differences in judicial handling
  among juvenile offenders of differing ethnicities

Box 2.2
(continued)

Percentage of births that are low birth weight
Asthma hospitalization rate for children
Voter participation in off-year primary elections
Library and community center usage
Public participation in the arts
Gardening activity
Neighborliness as measured by reported interactions with neighbors in
    community surveys
Perceived quality of life as measured by surveys of individuals' sense of
    well-being

Source: Sustainable Seattle, 1995. Indicators of Sustainable Community
1995.

be entirely warranted in specific cities because some cities may have a much greater need to account for pollutants than others. Additionally, some cities may be in the position where they can easily tap into existing sources of information from other places, such as a state agency, while others may not have that option. So one city's decision to include an indicator of air pollution and the absence of such an indicator in another city may simply be a reflection the kind of information available to those two respective cities. Indeed, sorting out what indicators are most appropriate and most readily available for which cities is one of the most challenging tasks of any effort to compare the seriousness of sustainable communities initiatives.

Although efforts to measure, and work toward reducing, the ecological footprint of a city must include concern for a variety of characteristics of the physical environment, sustainable indicators projects in cities have sometimes opted to consider a much broader array of social, economic, and political characteristics as well. As discussed next, it may be difficult for residents of a city to constrain their concern for sustainability to air pollution when their crime rates are soaring or when there are serious housing problems. How, one might ask, can cities with high crime rates claim to be working toward sustainability by focusing just on traditionally defined environmental pollution? If a city's housing stock is in serious decline, or if high costs of housing drive low- and modest-income residents out of the city, is it reasonable to expect that city to confine its sustainability efforts to those related to pollution? The same logic applies to a wide array of social and political issues that may seem somewhat detached from the ecological footprint definitions of sustainability.

Although all indicators of sustainability contain implicit value content, the development of indicators that stray from ecological, natural resources, and biophysical considerations present particularly salient opportunities for the value bases of sustainability to emerge. As an example, one review of the goals required to make the transition to sustainability suggests that there need to be "gender transitions away from male domination of public policy," and governance that is designed to "... promote freedom and social justice." (Corson 1993, 3) Strictly speaking, these are not measurement issues, *per se*, but rather issues of how sustainable communities are defined conceptually. Regardless of whether one embraces or rejects these particular notions, they nonetheless make much more visible the value-laden nature of some efforts to define sustainable communities.

Virtually all of the work that has been done on sustainable communities indicators recognizes the concept of sustainability as multidimensional, and that measuring urban sustainability therefore requires multiple indicators. Unfortunately, virtually all of the sustainable indicators projects have lacked extensive discussion concerning why specific indicators were selected, or more specifically, what aspect of sustainability they hope to measure. In other words, such projects are typically long on measures and short on rationales. Consequently, it is difficult to know the intent that lies behind the selection of any one or any combination of indicators. What this means is that, analytically, the task of understanding the various indicators that have been developed becomes a matter of inference. So, before embarking on an explication of these indicators themselves, it is important to understand the basic conceptual frameworks of the indicators that have been developed. After explaining the ways that indicators have been categorized, an examination of the indicators that flow from these categories will serve as the foundation for a discussion of their connection to "sustainability."

Perhaps the most elaborate articulation of sustainability indicators is found in the works of Walter Corson (1992; 1993) and of Elizabeth Kline (1995a; 1995b). Additionally, a comprehensive presentation of sustainable communities indicators has been produced by the Environmental Policy Center in San Francisco (Global Cities Online 1998). These broad reviews and comparisons of indicators used in various places demonstrate how much variation there is in the kinds of results sustainability projects might try to attain, and, in the process, how ambiguous the operationalization of the concept of sustainability is.

Corson, whose efforts were associated with the nonprofit Sustainable Seattle project, and initiated under the auspices of the Global Tomorrow Coalition, suggests that urban sustainability indicators fall into five fairly distinctly different subgroups: *ecological, environmental, natural resources issues; local economic performance and economic equity; ethical considerations; social–cultural issues;* and *political–governmental functions.* The ecological, environmental, and natural resources issues would appear to be those that are most directly related to Rees' ecological footprint conception. They involve the consumption of natural resources, and human activities that produce pollution. Cities might choose from among many dozens of ecological and environmental indicators, including the per capita number of gallons of gasoline consumed, amount of per capita electricity consumed, per capita volume of solid or hazardous waste produced, per capita air pollution emissions, and so on. Indicators of the function of the local economy and economic equity might include the unemployment rate, per capita income, indexes of income inequality, the percent of families living in poverty, the ratio of business start-ups to business failures, the ratio of the value of durable goods sold to services sold, and many others.

Sustainability indicators focusing on ethical, social, and cultural issues include the number of homeless residents, the number of violent crimes per 100,000 population, the overall crime rate, the frequency of child abuse and neglect, and others. Political–governmental function indicators include indicators of political participation, such as the percent of the population that is registered to vote and that votes, the frequency with which residents contact government officials, the incidence of volunteerism, the percentage of the population that is aware of the names of city councilors, and others. Governmental function indicators also include measures of governmental effectiveness, such as per capita cost of governmental services, and resident population ratings of governmental responsiveness, among many others.

Elizabeth Kline (1995a; 1995b), whose work was conducted in the context of the Cambridge (Mass.) Civic Forum under the auspices of the Consortium for Regional Sustainability based at Tufts University, suggests that categories of indicators represent "new paradigm" ways of thinking about communities. She examines indicators of the new paradigm *economic security* rather than the old paradigms of economic growth or economic development, *ecological integrity* instead of environmental protection, and the *quality of life for individuals and communities* rather than improved availability and delivery of social services. Without specific reference to any sort of old paradigm, she also focuses on *empowerment with responsibility* as an important category of sustainability indicators.

According to Kline's classifications, *economic security* indicators are concerned principally with reducing the creation of disparities in the function of the local economy, and with seeking the "environmentally sound utilization of natural systems." Economic security itself is measured by such indicators as the "hours of paid employment by household at average wage to support basic needs," the ratio of loans for micro versus large scale businesses, the percentage unemployed for more than one year (by race, gender, neighborhood), the creation of local wealth, and others. Environmentally sound utilization of natural systems is measured by a variety of "ecological integrity" indicators, including the "percent of energy used in a community generated by facilities using renewable energy sources," per capita or per household reduction and prevention of toxic waste streams brought into or leaving the community, extent of replacement of virgin materials with recycled products by households, business, and public institutions, to name a few.

Ecological integrity itself also refers to the "effectiveness of functional capacity of natural systems, as indicated by the value of damage from natural disasters, percentage loss of rare and endangered species, the percentage of drinking water that is lost or endangered, the percentage of water bodies meeting water quality standards, and so on. *Quality of life* indicators, coupled with empowerment, are really about "community building." These indicators focus on: self-respect and self-esteem, as in the presence and effectiveness of programs and activities to help people understand and respect peoples' differences, the presence of ethnic restaurants, and the availability of public communications in languages other than English; "caring," as represented by the frequency with which residents volunteer to help others; connectedness, as indicated by the prevalence with which residents know each other, the extent of interethnic personal interactions; and "basic coverage" or accessibility to human services, as represented by the extent to which there is equity in child care, quality housing, and security from crime; among many others. *Empowerment with responsibility* captures other elements of community building by focusing on the extent to which community residents are, and have access to opportunities to become, self-sufficient. This includes "reaching in" or outreach activities, such as mobilization of resident participation in civic life, accessibility to public meetings, open-ended solicitation of residents' views on proposed decisions, and others. It also includes creating an "equitable and fair playing field" for residents, and providing residents with the personal capacity to become engaged in public and civic life. "Responsibility" focuses on establishing accountability among residents and public officials, particularly for understanding the implications of their actions (or lack thereof) for others in the city. For example, it includes ensuring that community-based loans are repaid, that city budgets are goal directed, and that public employees reside in the city. Chapter 5 looks much more closely at the issues related to community building.

As innovative as this new paradigm of indicators might be, there do not appear to be any cities that have systematically adopted them. The indicators typically used in cities appear to be related more to the traditional or functional areas of city services. If the idea behind the new paradigm indicators was, at least in part, to usher in a new way of thinking about delivering and performing city services and functions, this certainly has not materialized.

Regardless of the terminology or the level of abstraction that is applied to the concepts that underlie these indicators initiatives, it is clear that many cities have come to the conclusion that to become more sustainable, they need to pay close attention to four main types of issues: ecological–environmental–natural resource issues; the performance of the local economy; a variety of quality of life issues; and long-term governance issues. Additionally, in one way or another, many of these initiatives consider social justice—equity and fairness—as important components of what it means to be sustainable. Examples of the indicators themselves may provide additional substance to these conceptual frameworks.

Indicators of the Environment, Ecological Health, and Ecological Footprints

As might be expected, by far the most commonly found, and perhaps most important, sustainability indicators focus on environmental issues. Indicators of the environment, ecological health, and ecological footprints, to use Rees's term, focus on aspects of the biophysical environment, environmental pollution and protection, natural resource use, preservation, and consumption, and a variety of other factors related to "ecological integrity." Indicators of these factors appear, at least on their face, to be the most directly related to environmentally oriented definitions of sustainability. As a consequence, indicators of the environment are discussed much more fully in chapter 3.

For many, if not most, of the environmentally oriented indicators, it is fairly easy to specify which direction is indicative of greater sustainability. For example, the Sustainable Seattle project included measures of resource consumption including the per capita consumption of renewable versus nonrenewable energy, the per capita number of gallons of gasoline bought, per capita gallons of water used, and the percent of locally produced food consumed. Presumably, greater consumption of nonrenewable energy resources reflects movement away from sustainability, while increased consumption of locally produced food reflects progress toward greater sustainability. Other communities' indicators initiatives have included per capita amounts of solid waste generated, hazardous waste generated, the percent of solid waste that is recycled, the percent of yard waste that is composted, the price of energy (electricity, natural gas, etc.) relative to personal income (a measure of how much market incentive there is for conserving energy), and the proportion of urban land used in agriculture.

Ecological footprint indicators also focus on the quality of the natural environment, mainly the air, water, and land. Air quality indicators include the amount of emissions of major pollutants by each sector of the local economy (local manufacturing industries, the energy sector, transportation, etc.), the amount of pollution in the ambient air regardless of sources, the number and types of toxic release incidents, and the acidity of precipitation. Water quality measures focus on the amount of pollutants released by each sector of the local economy, the quality and capacity of wastewater treatment facilities, the extent of contamination of lakes, streams, and aquifers, and the pollution content of drinking water. Land indicators focus on the number of Superfund hazardous waste sites, the presence of specific contaminants (lead, asbestos, etc.) in the soil at non-Superfund sites, the proportion of locally generated solid and hazardous waste that is disposed of within the community or shipped outside of the community, and the number of brownfield sites.

Indicators of Energy Consumption

Largely because of the linkage between the pollution that results from burning fossil fuels or generating electricity with nuclear generating facilities, indicators projects often develop measures of energy consumption. The expectation is that in order to become more sustainable, cities have to find ways of encouraging less use of polluting energy sources, and greater use of renewable energy sources. As discussed in more depth in chapter 3, many cities have begun modest programs to encourage the purchase of electricity that is generated from renewable sources, such as solar, wind, biogas, and other sources. The indicators of energy consumption usually focus on the amount of conventional energy resources that are consumed, where less is better from a sustainability perspective, or the amount of renewable energy resources that are consumed, where more is better. For example, in Scottsdale's very comprehensive indicators report, an effort is made to measure the number of kilowatt hours of electricity generated from solar sources that were used in the city, as well as the proportion of energy consumed that derives from solar. For this calculation, natural gas consumption is converted into kilowatt hours and combined with the measure of electricity (City of Scottsdale 2000). In Jacksonville, an indicator focuses on the amount of gasoline per capita that is consumed annually.

Indicators of Local Economic Performance

By now, it is clearly understood that the health and vitality of cities depends a great deal on the functioning of the local economy.

Sustainability approaches to local economic development, however, tend to be very different from mainstream "more development is better" approaches. Many cities have come to recognize that some kinds of economic development, often referred to as "smart growth," are better than others in terms of contributing to sustainability. For example, as is discussed in later chapters, particularly chapter 4, cities such as Pittsburgh, Pennsylvania, Chattanooga, Tennessee, and even Brownsville, Texas, have innovated in the pursuit of economic development by creating eco-industrial parks. The chief idea behind eco-industrial parks is to control and minimize the exportation of solid and hazardous waste from industries located in the city. Locating in close proximity companies whose wastes can be used by other local companies accomplishes this purpose.

Such operations can contribute greatly to achieving economic growth while minimizing the stream of solid or hazardous materials leaving the city. Most sustainability indicators projects, however, seem not to focus on the achievement of economic growth without sacrificing the environment per se. Rather, indicators usually concentrate on standard measures of economic performance. For example, Corson's "economy and economic equity" indicators include the unemployment rate, concentration of employment in a small number of employers (which, presumably is less sustainable than if employment is dispersed among a large number of employers), the percentage of families and individuals living below the poverty line, per capita income, the cost of living index (Consumer Price Index), municipal debt per capita, municipal bond ratings, and measures of the concentration or dispersion of personal income (Corson 1993, 13). Corson does list a handful of economic performance indicators that appear more directly related to the understanding the balance between economic growth and the environment. For example, the suggested indicators of "ratio of trade based on renewable resources to trade based on extractive or nonrenewable resources," and "the percent of local taxes designated as 'green taxes' on products that cause pollution, waste, or other environmental degradation" begin to capture aspects of the balance that sustainable cities may wish to strike. But most of the economic performance indicators can only capture the extent of the balance when viewed in combination with environmental indicators.

For Kline, many of the indicators of economic security are designed to more directly capture aspects of this balance, or at least to capture the environment side of the equation. For example, a major element of economic security, as noted previously, is "environmentally-sound utilization of natural resources" (Kline 1995a, 9–10). Indicators of this include the percentage of energy used in the community generated by facilities using renewable energy sources, the percentage of locally generated waste converted into beneficial uses (e.g., sludge to fertilizer pellets, waste steam to residential heating), the percentage of "raw materials" imported into the city that are recycled, and many others. But clearly, these indicators of "economic security" appear to have more to do with ecological or environmental security, although at some level the two are inextricably linked. It is nonetheless possible to imagine a situation where cities that track well on these indicators might experience declines in the more direct or traditional economic performance indicators.

Indicators of Sustainable Governance

While the exact manifestations of concern for sustainable governance vary from city to city, virtually all cities are concerned about the health of their "civil society." There is evidence that, for a variety of reasons, civic engagement in cities has declined (Putnam 2000), so sometimes the character of the local civil society is a target for sustainable indicators. However, not all cities that have developed sustainability indicators choose to include efforts to establish indicators of governance. The idea that there is some connection between issues of local governance and sustainability is one that is addressed more fully in chapter 5. Even among those cities that do include such indicators, the measurements tend to be rather crude in the sense that they probably tell very little about the overall health of the local civil society or local governance. For example, the Seattle Comprehensive Plan includes a variety of such indicators, goals, and actions in its "Human Development Element." It specifies goals to: make Seattle a place where people are involved in community and neighborhood life, where they help each other and contribute to the vitality of the city; strive to reach people in new ways to encourage broad participation in neighborhood and community activities and events; and promote volunteerism and community service; strengthen efforts to involve people in the planning and decision making that affect their lives (City of Seattle 2000).

Another example comes from Austin, Texas, where the Central Texas Indicators Project (2000) includes efforts to measure the extent of "civic engagement." Although civic engagement is a broad concept encompassing a wide array of activities in civil society, this indicators project focuses on two areas: the extent of participation (voting) in local elections; and support for charitable organizations. The single indicator of voter participation is the percentage of registered voters who voted in local elections, and the two indicators of support for charitable organizations are the percentage of residents who spent five or more hours doing volunteer work during a two-month period, and the percentage of residents who made financial contributions of at least $100 to charities (Central Texas Indicators 2000, 2000, 18, 22).

A third example, perhaps the best developed set of indicators related to issues of governance, comes from Jacksonville, Florida, which has developed seven "Indicators of Government/Politics." These include: the percentage of people who rate the quality of local government leadership "good" or "excellent" in an annual telephone-based public opinion survey; the percentage of the population 18 years old and over who are registered to vote; the percentage of registered voters who vote in general elections; the percentage of elected officials who are people of color or female; the percentage of people (in the annual opinion survey) who can name two current city council members; the percentage of people surveyed who report keeping up with local government news "frequently"; and the percentage of people surveyed who feel that local public services are effectively provided "frequently" (Jacksonville Community Council, Inc. 1999).

A few cities try to take a somewhat broader view of governance, and place it within the larger context of the health of the local civil society. In Santa Barbara, for example, two indicators focus on aspects of civil society. The first focuses on the total amount of money collected by the Santa Barbara United Way in its annual campaign, with the idea that increased contributions represent a healthier civil society. The 1999 indicators report suggests that total contributions increased substantially from 1961 through 1981, then fell through the early 1990s, then began to rise again. The second charts trends in local voter turnout in Presidential elections. Here the indicator shows that voter turnout from 1940 through 1996 declined rather sharply, paralleling the experience of most of the nation (Santa Barbara 1999).

Indicators of Equity and Equality

Issues of equity and equality are frequently articulated as major concerns in cities. As discussed in some detail in chapter 6, equity concerns in the context of environmental problems and related health effects have made their way onto the public agenda. Environmental equity and environmental justice generally relate to the differential ways that people of color, racial and ethnic minorities, and poor people are affected by environmental impacts. But a much broader range of equity and social justice concerns have made their way into sustainable indicators in some cities. Sometimes such concerns arise over how particular people or areas of a city are treated in the provision and delivery of public services, and sometimes such concerns arise over the more fundamental issues, such as the maldistribution of economic resources or other factors that affect the quality of life.

Although most cities that have sustainability indicators do not explicitly use indicators of social or environmental equity, some do. For example, San Francisco has developed three major indicators of social and environmental justice. These include the mean income level of residents of historically disadvantaged communities; the proportion of environmental pollution sources in historically disadvantaged neighborhoods compared to other neighborhoods; and the level of political participation of disadvantaged communities. In Jacksonville, Florida, efforts have been made to develop two indicators, one representing the proportion of the population that thinks racism is a problem in the city, and the other focusing on the number of employment discrimination complaints filed.

Actions Prescribed by Indicators

The development of sustainability indicators, by itself, does not suggest that there will be any particular achievement of desired sustainability results. For example, just because a city decides to periodically monitor and measure its level of air pollutants does nothing to work toward reducing those pollutants. So the process of developing sustainability indicators may include, or be coupled with, the specification of associated goals and actions that the community can or should take to achieve the goals. Sometimes the goals are simply to achieve more (or less) of whatever an indicator is measuring (e.g., to reduce sulfur dioxide levels in the air to levels lower than at some earlier point in time), or to achieve specific targets (e.g., to increase the number of days that the Air Quality Index is in the "good" range to 325 in the year 2002). Actions would specify what interventions the community needs to embark on to achieve the targeted goal. For example, a community might articulate an action of encouraging 20% of homeowners to convert from fuel oil to natural gas furnaces within a two-year period, and consequently adopt a specific program to help make that happen.

Clearly, a city that is able to articulate clear actions and interventions can be said to be more serious about achieving sustainability than one that does not, but prescribing actions appears to be the hard part. Among the many communities that have developed indicators, far fewer of them have begun specifying what kinds of actions will be necessary to achieve desired results as measured by these indicators. Indeed, many cities have found that the availability of information—basic data—that can be used to create benchmarks is not available, or may require significant expenditure of time and effort to develop. Nevertheless, some cities have made valiant efforts to prescribe actions and interventions in pursuit of their sustainability goals. Seattle's Comprehensive Plan sets many environmental and related goals, identifies the existing legal authorities related to these goals, and specifies what laws or other authorities would have to be changed in order to meet the goals. Those cities that identify and specify such actions certainly would seem to take sustainability more seriously than others.

Not all indicators are easily susceptible to prescribed actions. In other words, it might not always be clear what a city should do to improve the condition that the indicator purports to measure. Partly because the city may not possess the legal authority to pursue a particular sustainable course of action, and partly because there may be no known way to definitively achieve a particular goal, it is not always clear what a city's sustainability initiative can do to contribute to becoming more sustainable. While a particular city may fairly easily be able to prescribe actions to be taken to reduce the amount of incinerated solid waste, for example, it is much more difficult for it to prescribe ways of reducing the perception of racism among the population. Taking the actions may not be easy or cheap, but at least it is possible for a city to conceive of a way to make improvements. A city can take specific actions to make land use more environmentally friendly, but it would find few options available for stimulating greater voter turnout or charitable giving. Even in environmental issues, a city might find it difficult to know how to redress increasing problems with air pollution or energy consumption. What can a city do on its own to cut down on air pollution, particularly if the source of pollutants is found outside of the city? Yet the articulation of prescribed actions designed to meet goals specified by sustainability indicators represents an element in any serious sustainability initiative. In general, cities that make an effort to prescribe actions to implement elements of their sustainability plans would appear to take sustainability more seriously than cities that do not. In short, a sustainability initiative that combines an effort to measure progress with the articulation of specific goals or targets and prescribed programs and actions to achieve those targets would seem to be more serious about sustainability than those initiatives that lack these elements.

The Process of Developing Indicators

Almost as important as the indicators themselves is the process that is used in a given city to develop those indicators. Clearly much of the literature on indicators carries with it the implication or the assertion that some types of processes are superior to others. As noted previously, some cities have even made elements of their processes into explicit indicators of sustainability. When Kline (1995) suggests that one of the explicit purposes of indicators projects is "democracy building," where grassroots collaboration engages people in a community-building process, she is asserting an important role for the process. Indeed, imbedded in much of the work on sustainable indicators in specific cities is a sense that one of the unwritten purposes of these indicators projects is to promote ways of engaging residents in the process as part of such a community-building process. Moreover, engaging the participation of many "stake-holders" in the process is often considered to be an important, if not *the* important, determinant of how widely accepted the indicators become. As Zachary (1995) states it:

Based on the experiences of Seattle and Cambridge, it is evident that community involvement can be a key factor in developing tools for moving toward a more sustainable community. Without it, an indicators project may not receive wide acceptance and neglect to identify issues that are important to the vision of sustainability for certain segments of the community. (p. 30)

But there is not universal agreement concerning the need for broad-based community involvement in the development of indicators. Advocacy of such grassroots processes raises the all-too-frequent debate concerning the role of citizens in making decisions about highly technical issues, and the ability of any democratic process to make "good" policy decisions when scientific and technical expertise is required. Advocates of participatory processes seem to treat the concept of sustainability merely as a social construct, not as a concept that has objective meaning (perhaps even the same meaning from place to place). In his critique of the development of indicators in Seattle, Brugmann (1997b) suggests that the meaning of a sustainable city must be treated as something rather objective if it is to be treated seriously at all. In his words,

The Sustainable Seattle approach suffers from a common notion that the measurement of 'sustainability' can be achieved through a public participation process. . . . [but there is] tension in this notion. . . . The tension between scientific rigour and public values and perceptions . . . which arises from the ambiguity of the sustainability concept itself—. . . compromised the achievement of Sustainable Seattle's . . . objective [of developing indicators that integrate economic, social, and environmental phenomena]. If Seattle's sustainability is in fact a complex, ecologically determined condition, then the ability of Sustainable Seattle's dedicated generalists (i.e., stakeholders) to define and apply indicators to evaluate this condition accurately would appear problematic. (pp. 62–63)

So, to Brugmann, the key issue in the development of indicators is whether those indicators bear some relationship to the complex, objectively defined, ecological condition of the city, not whether there is broad social acceptance of some socially constructed definition that may have no relation to the city's ecology.

This, of course, raises a major dilemma in trying to ascertain w a city is taking sustainability seriously. The contrast could not be : As summarized next, there are two distinctly different, even contradictory, views on what makes an indicators project a serious action. The issue focuses on whether grassroots initiatives, particularly those that gain their impetus through efforts that are at least initially independent of city government, are more serious than those that are highly professional and oriented around defining very technical measures of ecological integrity. Table 2.2 provides a brief outline of arguments concerning the appropriate processes that should be used to develop sustainability indicators.

Although Brugmann's description ties the nature of public involvement to the nature of the resulting indicators' complexity (in this view, public involvement necessarily prevents development of complex indicators), clearly his concern is that for indicators to be part of a serious sustainability effort, those indicators must be directly and clearly related to what he calls "a complex, ecologically determined condition" of the city. Presumably, if some highly participatory process were able to accomplish this, Brugmann would have little difficulty accepting the process. His point, however, is that based on the experiences of the places he studied, participatory processes cannot produce the kinds of indicators he sees as necessary. Thus, while it might be useful to make distinctions among cities in their use of indicators, it is quite difficult to do so without making judgments about who is correct. A determination of who is correct may come eventually, but since each position sees its approach

Table 2.2
Appropriate processes for developing Sustainability Indicators

Advocate of this position	If the indicators are basic indicators, developed through grassroots effort	If the indicators are complex indicators, developed through professional expertise
Kline (1995a) Zachary (1995) MacLaren (1996)	More serious effort	Less serious effort
Brugmann (1997b)	Less serious effort	More serious effort

as being better able to lead to actual progress toward achieving sustainability, we may not know for many years which position is more efficacious. Not until we can independently measure and track sustainability over time, and correlate it with specific process characteristics, are we likely to know with any certainty, which position is correct.

Beyond Plans and Indicators: The Need for Coordinated Action

The creation of a sustainability plan, the establishment of an "indicators project" and the presence and use of the indicators is certainly an important element in determining how seriously a city seems to take sustainability. However, most observers seem to agree that having indicators, per se, is not enough. Of course the indicators must bear some close relationship to the actual achievement of improvements in the environment or quality of life for most, if not all, city residents. But even in the presence of such indicators, cities' efforts could quite easily fail to make much progress toward achieving their goals.

Perhaps the key test of the seriousness of issues of sustainability is what kind of integrated vision the city's sustainability plan carries, and whether the plan proposes to implement sustainability by incorporating the activities and responsibilities of numerous government, nonprofit, and business organizations. In his rather scathing critique of the nonprofit Sustainable Seattle, Brugmann argues that sustainability plans that lack truly coordinated visions are doomed to be able to perform only public education functions. In his view, such initiatives can never hope to make progress toward achieving actual sustainability results because they lack an understanding of any relationship between what they propose to measure in their indicators and how behavioral change is created. Brugmann suggests that the indicators developed in the Sustainable Seattle project, often cited as one of the very best such efforts in the country, accomplished much that was good:

But Sustainable Seattle itself, organized as it was without connection to major institutions, generally, and the City's strategic and statutory planning processes, specifically, neither provided a blueprint nor stimulated commitments, nor even a consensus, for action. Its impact in driving change in local conditions was therefore, at best, catalytic. (Brugmann 1997b, 64)

So, to Brugmann, if the goal of a city's sustainability initiative is to actually improve the biophysical environment, as opposed to raising public consciousness or conducting a public relations campaign, achieving a high degree of functional integration among local organizations and institutions is necessary. To take sustainability seriously, a city must incorporate the pursuit of progress toward improving the biophysical environment, as measure by sophisticated indicators, into its policies and planning processes.

To what degree do cities with sustainability plans or indicators projects actually achieve some high degree of functional integration or centralization? Most cities have not been particularly successful in finding administrative arrangements that seem to accomplish this goal. Yet a few have been able to place responsibility for sustainability in the city agencies that are able to perform coordinating functions. For example, Austin, Texas, operates its sustainability initiatives largely through its department of Planning, Environmental, and Conservation Services. Boulder, Colorado, established a department of Environmental Affairs that administers specific programs and coordinates many others. Seattle, Washington, has an Office of Sustainability and Environment, and Portland, Oregon, has an office of Sustainable Development. Many other cities with established sustainability initiatives elect to administer their efforts in a more piecemeal fashion, with responsibility dispersed throughout other functional city agencies. For example, in Jacksonville, Florida, the livable cities initiative does not fall to the responsibility of any specific agency. This is also the case in Chattanooga. And in still other cities, there is little manifestation of sustainability in city agencies at all, with whatever identifiable responsibility falling to various nonprofit organizations. In Olympia, Washington, for example, there is little manifestation of the initiative beyond that found in the Sustainability Roundtable, the nonprofit organization that organized the initiative.

What is perhaps a more challenging question is whether cities need to address issues of quality of life, social justice, and governance in order to be serious about achieving sustainability. Of course, the answer to this depends on whose definition of sustainability a city decides to adopt. Beyond this, however, the issue is one of whether and to what extent these elements of sustainability that are not directly linked to ecological and environmental concerns represent important, even necessary, conditions for cities to become sustainable. And answers to these questions push the issue of sustainable communities squarely into issues of contemporary political ideology. These nonenvironmental components of sustainability indicators are discussed more fully in the chapters to come.

An Index of "Taking Sustainability Seriously"

Based on the previous discussions, an index of sustainability can be developed to provide some basic sense of how cities compare. The comparative empirical analysis of sustainable cities requires some method of measuring the extent to which cities seem to take sustainability seriously. Although at some point it might be desirable and possible to determine how sustainable cities have actually become, the assumption of this analysis is that such assessments are premature. The analysis here, by contrast, focuses on the policies, programs, and activities of cities, discussed previously, that would seem to be consistent with an overall effort for cities to become more sustainable. In order to capture the essence of cities' efforts, the analysis here focuses on whether they have adopted or engaged in some thirty-four different specific activities in seven different categories, as summarized in table 2.3.

Although there are perhaps 60 or more U.S. municipalities that have been identified in the extant literature as having some form of sustainability initiative, as presented in table 1.1, the index is calculated for only twenty-four of these whose programs were in place as a matter of public policy as of January 1, 2000. Five of the listed municipalities are very small in population size (Grantsville, Utah; Ithaca, New York; Burlington, Vermont; Annapolis, Maryland; and Stuart, Florida) and were omitted from this analysis for that reason. The Lansing/East Lansing initiative was omitted because of its relative newness, having been started only in early 2001. There appears to be little tangible evidence of Oklahoma City's initiative, and lacking information about it, this city's program was also omitted. So the following analysis centers on identifying the programs, policies, and actions in these twenty-four cities.

The Elements of "Taking Sustainable Cities Seriously"

• The preceding discussion clearly suggests that one of the most frequently cited elements in any sustainable cities initiative is the sustain-

Table 2.3
The overall elements of the Taking Sustainable Cities Seriously index

Sustainable Indicators project
  1. Indicators project active in last five years
  2. Indicators progress report in last five years
  3. Does indicators project include "action plan" of policies/programs?

"Smart Growth" activities
  4. Eco-industrial park development
  5. Cluster or targeted economic development
  6. Ecovillage project or program
  7. Brownfield redevelopment (project or pilot project)

Land use planning programs, policies, and zoning
  8. Zoning used to delineate environmentally sensitive growth areas
  9. Comprehensive land use plan that includes environmental issues
  10. Tax incentives for environmentally friendly development

Transportation planning programs and policies
  11. Operation of inner-city public transit (buses and/or trains)
  12. Limits on downtown parking spaces
  13. Car pool lanes (diamond lanes)
  14. Alternatively fueled city vehicle program
  15. Bicycle ridership program

Pollution prevention and reduction efforts
  16. Household solid waste recycling
  17. Industrial recycling
  18. Hazardous waste recycling
  19. Air pollution reduction program (i.e., VOC reduction)
  20. Recycled product purchasing by city government
  21. Superfund site remediation
  22. Asbestos abatement program
  23. Lead paint abatement program

Energy and resource conservation/Efficiency initiatives
  24. Green building program
  25. Renewable energy use by city government
  26. Energy conservation effort (other than Green building program)
  27. Alternative energy offered to consumers (solar, wind, biogas, etc.)
  28. Water conservation program

Organization/administration/management/coordination/governance
  29. Single governmental/nonprofit agency responsible for implementing sustainability
  30. Part of a city-wide comprehensive plan
  31. Involvement of city/county/metropolitan council
  32. Involvement of mayor or chief executive officer
  33. Involvement of the business community (e.g., Chamber of Commerce)
  34. General public involvement in sustainable cities initiative (public hearings, "visioning" process, neighborhood groups or associations, etc.)

able indicators project (AtKisson 1996; Zachary 1995). Sustainable indicators consist of efforts to devise specific measures of how sustainable the city is, to establish benchmarks, goals, and timetables for improvements, and to periodically assess progress toward achieving these improvements. Sometimes these indicators projects are developed within city government, usually by a planning agency, as in Portland, Oregon, and sometimes they are developed by independent nonprofit organizations, such as the Sustainable Community Roundtable in Olympia, Washington. In a few cities, these projects are the joint effort of a nonprofit and one or more city agencies, as in Seattle. The index focuses on three aspects of these projects: (1) does the city have such a project; (2) if so, has the project issued a progress report within the last five years; and if so, (3) does the indicators project contain an explicit "action plan" delineating the steps to be taken to achieve the specified goals within the desired time periods.

• The second category of activity focuses on cities' adoption of "smart growth" programs or policies (ICMA 2001). Smart growth, discussed more fully in chapter 4, simply refers to any of a number of programs designed to help the city manage growth to avoid or eliminate suburban sprawl, and to direct economic development and population growth in ways that minimize their impacts on the physical environment. Here the focus is on (4) whether the city has managed development by developing eco-industrial parks; (5) has committed to cluster or targeted economic development; (6) has established one or more ecovillages; and (7) has established a local brownfield redevelopment initiative.

• Central to the issue of sustainability is the broad issue of land use planning and the use of zoning. Increasingly, cities endeavor to use zoning to manage their growth, and just as with other smart growth initiatives, to try to take a comprehensive view of how the land in the city is developed into the future. As discussed in chapter 4, there is certainly a great deal of debate concerning whether the use of such land use controls represents an effective mechanism for achieving sustainability. Yet, the literature on sustainability is unambiguous about the need to change the ways that development takes place in cities. The focus here is on (8) whether the city engages in comprehensive land use planning that explicitly delineates environmentally sensitive growth areas; (9) whether the city uses zoning as a mechanism to influence the directions of development, for example, does the city's zoning ordinance establish environmentally sensitive areas; and (10) does the city attempt to use local tax incentives or other financial incentives, such as fee waivers, to influence development toward less environmentally sensitive areas.

• Partly as an extension of smart growth efforts and environmentally sensitive land use planning, as discussed in chapter 3, transportation planning has become an important element of sustainable cities initiatives (Newman and Kenworthy 1999). While much of what goes on in sustainable public transportation planning is captured in the "cluster development" approach to smart growth, it also includes other elements that are not. In the simplest case, cities can operate their own mass transit systems (buses or subways), and work to encourage more commuters to opt to use them. The focus here is on whether the city: (11) operates its own intra-city system of mass transit; (12) establishes limits on the availability of downtown parking spaces (creating incentives for commuters to seek means of transportation other than the personal automobile); (13) has defined car pool programs, including the use of car pool lanes on local roadways; (14) establishes a program for the city's fleet of vehicles to use alternative fuels (such as LNG, propane, or electric hybrid vehicles); and (15) operates a bicycle ridership program, with defined bicycle lanes and paths for commuters.

• Sustainability requires explicit attention to issues of pollution remediation, reduction, and prevention (Lachman 1997). As discussed in chapter 3, cities vary greatly in the extent to which they engage in activities, programs, or policies that are designed to address issues of pollution. The focus here is on whether cities have programs dedicated to: (16) household solid waste recycling; (17) industrial or commercial solid waste recycling; (18) hazardous waste recycling; (19) air pollution reduction, e.g., VOC reduction programs; (20) city purchasing of recycled products; (21) Superfund or other hazardous waste site remediation; (22) asbestos abatement; and (23) lead paint abatement.

• As discussed in chapter 3, Energy and Resource Conservation initiatives typically define programs to either reduce energy or resource consumption, or to change the forms of energy consumed to move away from the use of fossil fuels toward renewable energy sources. Even independent of initiatives associated with public transportation policies and programs, many cities have developed "green building" programs to assist developers with technical issues on design and construction of energy efficient buildings. In some cities, such as Boulder, specific green building elements must be included in the plans of new construction in order to obtain a building permit. The focus here is on whether the city has: (24) a green building program, either voluntary or mandatory; (25) a program of renewable energy use by city government; (26) any type of energy conservation program other than that found in a green building initiative; (27) made provisions for residential consumers to purchase electricity generated from alternative renewable sources (solar, wind, biogas, etc.), as in Austin, Texas, and Santa Monica, California; and (28) established a water conservation initiative of any sort.

• Although most discussions of sustainable cities place little emphasis on organizational, administrative, and governance issues, increasingly these are treated as important elements in the seriousness of cities' efforts (Brugmann 1997b). A city that purports to operate a sustainability initiative, but where the responsibility for making progress toward sustainability is undefined or dispersed around the city, usually means that sustainability will be subordinated to some other administrative goals. A city that designates a single department or administrative unit whose success is directly determined by the success of the sustainability initiative would seem to take the issue more seriously. Additionally, the level of involvement of key policymakers in the process of defining the initiative reflects the level of commitment of the city. In Boston and San Francisco, for example, those who have worked on their respective sustainability programs lament the lack of involvement of their mayors, and attribute their limited successes to this factor. Additionally, much of the attention that has been generated by sustainability initiatives focuses on the "empowerment" and participatory potential for cities residents. As discussed in chapters 5 and 6, much of the impetus for pursuing sustainability programs comes from a desire to define a process that is perhaps more inclusive and egalitarian than is found in cities' regular policymaking processes. As a result, the focus here is on whether the city: (29) has a single governmental or nonprofit organization that is responsible for implementing the program; (30) has made sustainability part of its overall comprehensive management plan; (31) has involved members of the city council or planning council; (32) has involved the mayor or chief executive officer (such as a city manager) in the deliberations and development of the initiative; (33) has involved the business community, either through involvement of specific businesses or business leaders, or through the local chamber of commerce; and (34) has involved the general public in some fashion, through public hearings, the oft-practiced "visioning" process, or through engagement with existing neighborhood associations and organizations.

These thirty-four elements are summarized into a single index representing the total number of program elements practiced in each city. This index focuses just on the programmatic elements found in the cities, and it assumes that each element is equally important in contributing to taking sustainability seriously. One might argue that these elements are not equal, and that some are far more important than others. For example, a city that has only pursued sustainability within the context of comprehensive planning is taking sustainability far more seriously than one that has established a solid waste recycling program. Yet there is no way of knowing how these elements should be weighted, at least not in terms of their overall contribution to making the city more sustainable in fact. At some point in the future, we may be able to determine the relative importance of each element in creating greater sustainability, but until better data exist to measure sustainability directly, there may be no way to weight the index elements.

The computed index value for a specific city could theoretically range from 0 to 34, although the fact that a city has undertaken a sustainability initiative makes a 0 score highly unlikely among the twenty-four cities examined here. The cities' actual index values range from 6 in Milwaukee to 30 in Seattle, as shown in table 2.4. This table also provides a full list of the elements and how each city was coded on the elements. The average index score is 16.6, with a standard deviation of 7.2. Clearly, some program elements appear to be easier to achieve than others. One element, solid waste recycling, exists in all twenty-four cities. Another element, water conservation programs, exists in twenty-three cities. On the other hand, the creation of ecovillages and the programs to operate car pools and car pool lanes on the roadways were each achieved only by three cities, respectively.

With all the caveats discussed earlier, the index provides at least a glimpse into the relative rankings across cities. There is a clearly identifiable cluster of cities, from Seattle (index score of 30) to San Francisco (score of 23) that also includes Scottsdale, San Jose, Boulder, Santa Monica, and Portland. These cities, perhaps precisely the ones that might be expected to be among those at the top of such a ranking, clearly are doing more than others to pursue sustainability. A second grouping, from Tampa to Boston, with scores from 14 through 19, respectively, contains some possible surprises. It is perhaps surprising that some cities, such as Boston and Cambridge, Massachusetts, are not in the top grouping, and other cities, such as Chattanooga, Tennessee; Austin, Texas; and Jacksonville, Florida, are relatively highly ranked. A third grouping, from Milwaukee to Orlando, with scores from 6 to 11, respectively, is clearly composed of cities that seem to take sustainability less seriously.

Table 2.4
Twenty-Four cities' scores on the "Taking Sustainable Cities Seriously" index

Sustainable Cities Initiative Program Elements*

City	Score	1	2	3	4	5	6	7	8	9	10	11	12	13	14	15	16	17
Seattle	30	Y	Y	Y	N	Y	Y	Y	Y	Y	Y	Y	N	Y	Y	Y	Y	N
Scottsdale	26	Y	Y	Y	N	Y	N	Y	Y	Y	N	Y	N	N	Y	Y	Y	Y
San Jose	26	Y	Y	Y	N	Y	N	Y	Y	Y	Y	Y	N	N	Y	Y	Y	Y
Boulder	26	Y	Y	Y	N	Y	N	N	Y	Y	N	Y	N	N	Y	Y	Y	Y
Santa Monica	25	Y	Y	Y	N	Y	N	Y	Y	Y	Y	Y	N	N	Y	Y	Y	Y
Portland	25	Y	Y	Y	Y	Y	N	Y	Y	Y	N	Y	N	N	Y	N	Y	Y
San Francisco	23	Y	N	N	N	Y	Y	Y	Y	N	N	Y	Y	Y	Y	Y	Y	Y
Tampa	19	Y	Y	Y	N	Y	N	Y	Y	Y	N	Y	N	N	Y	Y	Y	Y
Chattanooga	18	Y	N	N	Y	Y	N	Y	N	Y	N	Y	N	N	N	N	Y	Y
Tucson	18	Y	N	Y	N	Y	N	Y	Y	Y	N	Y	N	N	Y	Y	Y	Y
Austin	17	Y	Y	Y	N	Y	N	Y	Y	N	Y	Y	N	N	Y	Y	Y	N
Phoenix	15	N	N	N	N	Y	N	Y	N	N	N	Y	N	N	N	N	Y	N
Jacksonville	15	Y	Y	N	N	Y	N	Y	Y	Y	N	Y	N	N	Y	Y	Y	N
Cambridge	14	Y	N	N	N	Y	N	Y	Y	N	N	N	Y	N	N	N	Y	Y
Cleveland	14	Y	N	Y	Y	N	Y	Y	Y	Y	N	N	N	N	Y	Y	Y	N
Brookline	14	Y	N	N	N	Y	N	N	Y	Y	N	Y	Y	N	N	N	Y	Y
Boston	14	Y	Y	N	N	Y	N	N	Y	Y	N	Y	Y	N	Y	Y	Y	Y
Orlando	11	N	N	N	N	Y	N	N	N	N	N	Y	N	N	Y	N	Y	N
Santa Barbara	10	Y	Y	N	N	Y	N	N	Y	Y	N	Y	N	N	Y	Y	Y	Y
Indianapolis	9	N	N	N	N	N	N	Y	N	N	N	Y	N	N	N	N	Y	N
Olympia	8	Y	Y	Y	N	Y	N	N	N	N	N	Y	N	N	N	N	Y	N
New Haven	8	N	N	N	Y	Y	N	N	N	Y	N	N	N	N	N	N	Y	N
Brownsville	7	N	N	N	Y	Y	N	Y	N	N	N	N	N	N	N	N	Y	N
Milwaukee	6	N	N	N	N	N	N	Y	N	N	N	Y	Y	Y	N	N	Y	N
Number of cities	—	18	12	11	4	21	3	18	15	15	4	21	5	3	14	14	24	13

Table 2.4
(continued)

Sustainable Cities Initiative Program Elements*

City	18	19	20	21	22	23	24	25	26	27	28	29	30	31	32	33	34
Seattle	Y	Y	Y	Y	Y	Y	Y	Y	N	Y	Y	Y	Y	Y	Y	Y	Y
Scottsdale	Y	N	Y	Y	N	N	Y	Y	Y	Y	Y	Y	Y	Y	Y	Y	Y
San Jose	Y	Y	Y	N	N	N	Y	Y	Y	Y	Y	Y	N	Y	Y	Y	Y
Boulder	Y	Y	Y	N	Y	Y	Y	Y	Y	N	Y	Y	Y	Y	Y	Y	Y
Santa Monica	Y	N	Y	N	N	N	Y	Y	N	Y	Y	Y	Y	Y	Y	Y	Y
Portland	Y	Y	Y	Y	N	N	Y	Y	Y	N	Y	Y	Y	Y	Y	Y	Y
San Francisco	Y	N	Y	Y	Y	Y	Y	Y	Y	N	Y	Y	Y	N	N	N	Y
Tampa	N	N	N	N	N	N	N	Y	Y	N	Y	Y	Y	N	N	Y	Y
Chattanooga	Y	Y	N	Y	N	N	N	N	N	Y	Y	N	Y	N	N	Y	Y
Tucson	Y	N	N	N	N	N	Y	N	Y	N	Y	N	Y	Y	Y	Y	Y
Austin	N	Y	N	N	Y	N	N	Y	N	Y	Y	Y	Y	N	N	Y	N
Phoenix	Y	Y	Y	N	N	N	N	N	N	N	Y	Y	Y	N	N	N	N
Jacksonville	N	N	N	N	N	N	N	N	N	N	Y	Y	Y	N	N	N	Y
Cambridge	N	Y	Y	N	N	N	N	N	N	N	N	Y	Y	N	N	Y	Y
Cleveland	Y	N	N	N	N	Y	N	N	N	N	Y	N	N	Y	N	N	Y
Brookline	N	N	Y	Y	N	N	N	Y	N	N	Y	Y	Y	N	N	Y	Y
Boston	Y	N	N	Y	N	Y	N	N	N	N	Y	Y	Y	N	N	N	Y
Orlando	N	N	N	N	N	N	N	N	N	N	Y	N	N	N	N	N	Y
Santa Barbara	N	Y	N	N	N	N	N	N	N	N	Y	Y	N	N	N	N	Y
Indianapolis	Y	Y	N	N	Y	N	N	N	N	N	Y	Y	N	N	N	N	Y
Olympia	N	N	N	Y	N	N	N	N	N	N	Y	N	N	N	N	N	Y
New Haven	N	N	N	N	N	N	N	N	Y	N	Y	N	N	N	N	Y	Y
Brownsville	N	N	N	N	N	N	N	N	N	N	Y	N	N	N	N	Y	N
Milwaukee	N	N	N	N	N	N	N	N	N	N	Y	N	N	N	N	N	Y
Number of cities	13	10	11	6	5	5	8	10	7	6	23	16	14	9	7	13	21

* See table 2.3 for Initiative Program Element Listing.
Y signifies that the city has this program element; N signifies that the city does not have this program element.

These rankings not only serve to distinguish the cities from one another, they also lend themselves to great speculation about why some cities take sustainability so much more seriously than others. The rankings raise many questions concerning how one might explain this variation. If sustainability is seen as a progressive policy, then finding cities such as San Francisco, Seattle, Santa Monica, and Portland high on the list would not be particularly surprising. Yet the list demonstrates that (1) not all seemingly progressive cities rank high on the list; and (2) not all of the highly ranked cities would be thought of as being particularly progressive places. So this raises a variety of questions concerning what other factors might explain this variation. Are there characteristics that high-ranked cities share, and that lower-ranked cities do not share? This is an issue that is discussed in some detail in chapter 8.

Taking Sustainability Seriously?

How do we know whether a city is taking sustainability seriously? Can we look at one city and make a judgment concerning whether it takes sustainability more seriously than another city? Given the number of city-based initiatives, the array of indicators that have been developed, and the multidimensionality of the concept itself, and the differences in the processes used to develop indicators, this may seem like an insurmountable task. Yet, if it is desirable to compare cities to see if anything can be learned from their experiences, particularly in terms of trying to determine why some cities take sustainability more seriously than others, that is precisely what must be done.

The sustainable indicators initiatives provide an outline of what kinds of actions cities think they need to engage in if they wish to take sustainability seriously. To begin with, cities that have gone through the development of sustainability indicators, by virtue of this action alone, would appear to take sustainability seriously. These cities can be subdivided according to whether the indicators initiatives include comprehensive strategic planning elements, so that a city that outlines specific government actions, for example, to achieve specific goals can probably be said to be more serious than a city that does not. But does this necessarily mean that any city that has developed a sustainability indicators initiative takes sustainability more seriously than another city that has not? Although this is likely the case, one can imagine a situation where a city is engaged in an enormous array of efforts to become sustainable without making these efforts part of a larger, explicit, sustainable communities project. Although this might be rare, it is possible to imagine a city whose sustainability efforts are completely integrated with the everyday functioning of city government and its agencies, and as a consequence, no plan qua plan exists. Perhaps the closest example to this may be found in Portland, Oregon, where the city's planning department works under a comprehensive integrated conception of sustainability. At first glance, Portland would seem to not have a sustainability plan, per se. But upon further investigation, sustainability considerations flow out of almost every part of community and economic development planning there.

Perhaps equally important, sustainability indicators projects often provide limited utility in measuring the extent to which cities take sustainability seriously when the indicators they adopt are themselves only poor operationalizations of the concept of sustainability. For the most part, however, taken as a whole, cities' indicators capture a wide array of important areas where achieving results would indeed seem to constitute progress toward sustainability.

There are major differences of opinion concerning what constitutes a serious sustainability initiative or a serious indicators project. (Brugmann 1997a; 1997b; Pinfield 1997). One argument suggests that if a city is serious about measuring sustainability, it needs to develop a highly sophisticated and complex set of technical measures, particularly with respect to the environment. In this view, the process of developing indicators must necessarily be a professional enterprise, one conducted by experts who have the technical expertise to know what ought to be measured and how best to measure it. Another argument says that the process must be resident and neighborhood driven, responding to what the people in the city think is important. For advocates of this view, indicators of sustainability are completely value-driven social constructs rather than purely objective quantitative measures of universally important elements of environmental quality. Those who adhere to the latter view are not likely to think highly of seeing a city hire an outside consulting firm to apply a packaged set of indicators, as Tampa has done. On the other hand, those who see indicators as management and implementation tools might see such an effort as being an efficient step toward achieving sustainability if the packaged indicators are technically and professionally well done.

Still, the question remains: are there elements of a city's sustainability initiative that would appear to reflect a higher degree of seriousness about achieving results? The preceding discussion and the following chapters suggest that there are. As summarized briefly in table 2.4, each of the elements makes a contribution to the seriousness of the initiative. In the best of all worlds, we would like to be able to assign definitive weights to the importance of each element. But there is no acceptable methodology for doing so. For the purpose of this analysis, the judgments found in table 2.4 represent reasonable assumptions about cities' efforts. The role that each element plays in creating a serious sustainability initiative, and the rationale for the judgments concerning these roles, are explained in the chapters that follow.

One explicit area omitted in this discussion has to do with the budgets of local governments. Some might argue that unless cities are willing to dedicate significant amounts of funds to their sustainability initiatives, then the enterprise will be doomed. Perhaps more important, an argument can also be put forth that a city that spends more money on sustainability-related activities can be said to take sustainability more seriously than one that spends less. As compelling as this argument might be, and it will not be refuted here, the pragmatic problem is that comparisons of city budgets are extremely difficult to conduct. In its Census of Governments, the U.S. Census Bureau has not created a standardized spending category called "sustainability," or even "the environment," the way it has for police services, education, and other functional areas. Although individual city budgets are available, there is no standardized place to look to identify the amount of money spent on sustainability activities. Indeed, given the standard line-item budget used by most cities, it is often impossible to identify the programs and activities related to sustainability where city revenues have been budgeted or spent. It is for this reason that the assessment of "taking sustainability seriously" has opted to focus directly on programs, policies, and activities that have been defined as a matter of public policy. By focusing on these areas, this effort can begin to build a picture of how seriously cities seem to be about sustainability.

One last area considered but ultimately not included focused on whether the cities have policies or programs designed to reduce the use of lawn fertilizers and pesticides, frequently cited as sources of watershed contamination. This pollution prevention and reduction element was not included here because of difficulty obtaining definitive information about what the twenty-four cities were doing. Even with these omissions, the resulting Index still provides a reasonable assessment of how well cities seem to be doing in their official pursuit of sustainability.

3

The Environment, Energy, and Sustainable Cities

Without a doubt, the single most important element in any city's sustainability effort revolves around the environment, and by extension, energy usage and conservation. As discussed earlier, not all conceptions of sustainability or sustainable cities are limited to issues of the environment, but the environment plays a central role in virtually all of them. In what ways have cities attempted to address these kinds of issues? How have cities sought to protect and improve their biophysical environments, maintain ecological integrity, or affect energy usage within their borders? As noted earlier, cities do not always possess the wherewithal to affect these kinds of issues, yet within the jurisdiction of a typical U.S. city, there is plenty of environmentally relevant work to be done. The purpose of this chapter is to examine the range of activities that cities have engaged in to try to affect the quality of the environment. It will examine the targets of city-based initiatives, and highlight efforts at pollution reduction and prevention, remediation, and energy conservation.

Sustainable Cities and the Environment

Traditionally, cities have not been among the nation's leaders in pursuing environmental protection. The evolution of interest in, and ability to affect, the range of human activities that have been responsible for environmental degradation in cities probably began in earnest in the early 1970s. Since that time, however, cities all over the country have come to understand the importance of a clean environment to the quality of life and prospects for economic growth. As noted later, the city government in Chattanooga seemed to have no interest in preventing the extreme deterioration of the air quality there until the problem became so serious that it clearly affected the quality of life. The city of Cleveland showed no particular interest in protecting its water resources until the mid-1960s when the Cuyahoga River became so polluted that it repeatedly caught on fire (Liroff 1976, 3). Today, cities all around the country are much more attuned to the integral importance of a clean environment, and this is certainly reflected in the activities of cities that have elected to pursue some sort of sustainability initiative. Cities typically have programs that seek to protect ground and surface water, to recycle household and industrial solid waste, to reduce air emissions, to conserve energy and promote the use of cleaner sources of energy, and redevelopment of brownfields, to name a few. Additionally, cities often pursue economic development through encouraging environmentally friendly development, or what has become known as "smart development," an issue that is discussed more fully in chapter 4.

Focusing very specifically on the twenty-one environmental elements of sustainability initiatives represented by items eight through twenty-eight in table 2.2, it is clear that cities vary considerably. Although other elements may be related to the environment, these twenty-one elements represent very specifically targeted programs or activities that are directly designed to affect the quality of the environment. Other elements, particularly those related to "smartgrowth," (items 4 through 7) are examined more closely in chapter 4. By simply adding up the number of elements for each of the twenty-four cities, as listed in table 2.3, the resulting environmental score shows that Seattle ranks as having the most complete treatment of environmental issues, with a total of 18 of the 21 program elements. San Francisco ranks second, with 17. Olympia and Brownsville fall at the bottom, with 3 elements each. The average (mean) number of environmental program elements across all 24 cities is 9.7. Even within these scores, cities still vary considerably in the kinds of programs and activities they pursue; some specific case examples will illustrate this variance.

Water and Air Quality Efforts

Efforts in cities to improve and protect water and air quality often represent significant tests of the willingness of local governments to deal with transboundary issues. Because most cities do not have legal jurisdictions over the full range of water resources or over the facilities where air pollution originates, cities often find dealing the water and air quality issues to be a significant challenge. Whether the issue is wetlands protection, watershed management, aquifer protection, wastewater treatment, the management of recreational waterways, industrial or household air emissions, or many other water and air quality issues, cities often must look beyond their borders to seek some form of collaborative multijurisdictional effort. In a metropolitan area, there is substantial potential for these water and air resource challenges to become problems of commons resources. Suffice it to say here that sustainable cities engage in cooperative efforts to avoid creating commons problems and minimize transboundary impediments to sustainability. Additionally, because air and water quality issues often involve areas larger than the cities themselves, addressing these issues may also involve the integrity of surrounding ecosystems.

Air Quality   Actions to try to protect the quality of the air represent some of the more challenging aspects of environmental protection for cities. Cities often do not control the point sources of pollution, particularly when air emissions originate far downwind of the city. Non-point source emissions, such as vehicle exhaust emissions, are typically in the domain of state regulation, implemented through counties or air quality control districts, and cities have few options for directly affecting them. The vast majority of cities, including those that have active sustainable indicators programs, do not have explicit air pollution control or reduction programs, per se. As a consequence, they may not have associated air quality targets, goals, or even indicators. Cities' efforts to control air emissions often focus on indirect measures related to transportation or energy conservation, sometimes offering public transportation alternatives to reduce the number of vehicles driven or seeking to promote reduced fuel consumption. Numerous other initiatives have also been undertaken. The cities of Santa Monica, Seattle, Boulder, and Chattanooga serve as illustrations.

Santa Monica has gone about as far as any city in trying to affect its local air emissions particularly from sources generated by the city government itself. It is located near Los Angeles, an area that is widely known to have serious air pollution problems. The sustainability initiative in Santa Monica incorporates air emissions indicators as part of its effort to reduce energy consumption, and targets emissions of "greenhouse gases"—in this case, CO_2 and methane gas. Although the city has not adopted a specific goal or target for reductions of greenhouse gases, it has established a system for monitoring emissions from local sources, and tracks these emissions by sector (residential, commercial, industrial, transportation, and solid waste). As might be expected, over one third of the measured emissions come from transportation (mostly from automobiles), and another third originate with the commercial sector. From 1990 to 1997, the period covered by the last progress report, Santa Monica reports an overall 5.2% reduction in greenhouse gas emissions.

Although it is not possible to determine how much of the reduction has been produced explicitly by city-wide, or even specific city government, efforts, Santa Monica has taken numerous steps to try to reduce emissions. Most of these actions attempt to improve energy efficiency and the use of renewable energy sources, offer alternatives to single-passenger automobile trips, encourage alternative fuel vehicles, create incentives for greater recycling and waste reduction, and establishing a green building program. For example, the city has an aggressive program to retrofit city facilities with energy efficient heating and lighting equipment. It elected to purchase 100% of the electricity for city facilities from renewable sources. It replaced all city-owned or leased office equipment with energy efficient equipment. It installed photovoltaic electric generating systems in a number of city-owned facilities. And it has embarked on a project to retrofit the city's traffic signals with energy efficient lamps. Additionally, as part of its transportation plan, Santa Monica has sought to attain the explicit goals of increasing public transit ridership by 10%, increasing the average passengers per private vehicle to 1.5 from 1.0, and converting 75% of the city's fleet vehicles to use reduced-emission fuels.

Efforts to improve air quality in Seattle have been incorporated into both the Sustainable Seattle initiative and the Seattle Comprehensive Plan of the city government (City of Seattle 2000). Sustainable Seattle has focused on a single indicator, the number of good air quality days per year, where a good air quality day is defined as one where none of four air pollutants (carbon monoxide, sulfur dioxide, ozone, and suspended particulate matter) exceeds an established "good" standard for that pollutant (Zachary 1995, 13–14). The city government's Comprehensive Plan does not contain explicit indicators of air quality, but it does incorporate an air quality goal to "strive to reduce air pollution from all sources, including transportation, wood burning, and industrial activities" (Seattle Comprehensive Plan 2000). This goal is accompanied by a number of specific policies designed to address issues of growth management, increased use of public transit, the promotion of low- and zero-emission vehicles by large fleet operators, and aggressive identification and enforcement of state regulatory standards as applied to local industries.

Perhaps one of the more impressive local initiatives to reduce air pollution comes from Boulder, Colorado. Although Boulder does not have explicit air pollution targets among its sustainable indicators, it does operate two air pollution initiatives. One of these is a program run in conjunction with the Boulder County Clean Air Consortium, in which a variety of public education campaigns are sponsored to raise awareness about how individuals can reduce air emissions. For example, it sponsors the "don't top it off" effort designed to educate consumers about the emissions from gasoline stations, and an initiative to convince people to leave their cars at home one day a week. Additionally, it sponsors a voluntary reporting system for local industries to document the air emissions reductions they have achieved.

The second program in Boulder focuses on an effort to assist small businesses in finding ways of reducing their emissions. The Partners for a Clean Environment, or PACE, program, is a joint effort mainly involving the city of Boulder Office of Environmental Affairs, the Boulder County Health Department, the city of Longmont, the Boulder Energy Conservation Center, and the Boulder Chamber of Commerce. In order to work toward air emissions reductions, this program has targeted a number of types of small businesses for assistance in developing emissions reductions strategies. As the program notes, "the initial focus of the program has been smaller, generally unregulated businesses which collectively have a significant impact on public and environmental health. Smaller businesses often lack the knowledge and resources to investigate P2 [pollution prevention] alternatives. The auto repair, auto body, and printing sectors were targeted first because of the availability of P2 alternatives and the potential impact of P2 for these sectors" (PACE 2000).

These initial efforts focused on reductions of volatile organic compounds, or VOCs, from auto repair shops and from printers. The program essentially worked with business owners to make them aware of technologies that would allow them to conduct their businesses while polluting less and sometimes at lower operating costs. For example, the program disseminated information to auto paint shops about the availability of high-volume, low-pressure paint spray guns that produce lower emissions, and encouraged the use of nonhazardous detergents rather than solvents for cleanup. In the printing industry, the program encourages reductions in the use of isopropyl alcohol in printing solutions, and increased use of blanket and roller washes, and vegetable-based inks. According to the program, these efforts have reduced the volume of VOC emissions in Boulder by 25 tons per year.

Chattanooga, which has had a significant history of problems with air pollution stemming from its geographic proximity to mountain ranges and the proliferation of heavy industry through the 1960s and into the 1970s, has become a national leader in the use of alternatively fueled public transit vehicles. When the city elected to pursue the use of electric buses, it ended up facilitating the creation of a company dedicated to the manufacture of such vehicles. To some degree, this effort stands as a cornerstone not only of the city's effort to improve the quality of the air, but also as an example of sustainable economic development. Here is a case where the city sought to create economic development and add to its employment base in a way that contributed positively to improvement of its air quality. Not only does the resulting company, Advanced Vehicle Systems, Inc., produce electric and hybrid transit vehicles for the city, it also exports these buses to other cities. Today, this manufacturing company operates in a sort of symbiotic relationship with the Electric Transit Vehicle Institute, a nonprofit organization dedicated to promoting the design, production, and utilization of battery-powered electric and hybrid electric vehicles.

Water Quality   There is probably no single greater contributor to the health and well being of the population of a city than the quality of its water resources, particularly its sources of drinking water. Cities are required by federal law to take measures to protect their drinking water, but cities often take this challenge even more seriously. Although incidents of seriously contaminated drinking water are not particularly common, failure to protect the water of a large city can pose an enormous threat to the health of its people. As a result of the 1993 incident involving cryptosporidium, Milwaukee found out the hard way what happens when a city's primary source of water becomes massively contaminated. Cryptosporidium is a microscopic protozoan that, when ingested, causes diarrhea, fever and other gastrointestinal symptoms. The organism is found in many surface water sources and comes from animal wastes in the watershed. It can be managed or eliminated by a combination of water processing treatments including ozonation, coagulation, sedimentation, filtration, and disinfection. In Milwaukee, over 400,000 people became ill after drinking city water that had become contaminated. The problem was traced to the fact that the city uses Lake Michigan as both the source of drinking water and the recipient of treated sewage that, under certain conditions, flowed back toward Milwaukee's intake pipes (MacKenzie et al. 1994). Since that time, the Metropolitan Milwaukee Sewerage District has made significant changes to its operations in order to try to prevent future contamination.

All cities are required by federal and state laws to test their water supplies to ensure that drinking water supplied by the city meets regulatory standards for specific contaminants. Cities vary, of course, in how they get their drinking water and how vulnerable to contamination the supplies of water are. Cities also vary in terms of their access to surface and groundwater resources, including those available for recreational purposes. Yet nearly all cities involved in sustainability have made significant efforts to protect, conserve, and improve their water supplies and resources beyond what is required by federal or state law. The cities of Chattanooga and San Francisco serve as prime examples.

In Chattanooga, the sustainability effort includes significant attention to a wide array of water resources and issues. It includes, for example, coordinated efforts on protection of the Tennessee River Watershed, the North Chickamaugua Creek Gorge Watershed Protection Project, the Tennessee River Gorge Project, a stormwater management program, the Tennessee Riverpark Project, a regulation-oriented Groundwater Resources Protection Program, and the Chattanooga Creek Cleanup

Project. Taken together, these efforts represent a significant attempt to protect and improve the quality of water resources in and around the city.

The effort is certainly not devoid of problems, as illustrated by the Chattanooga Creek cleanup project. The Chattanooga Creek traverses through a heavily industrialized section of the city. The Creek runs 23.5 miles, east to west, from its origin in extreme northwestern Georgia. As a result of years of industrial waste disposal in the Creek, it has been designated as a National Priorities List Superfund site. The Creek's sediments are contaminated with a variety of hazardous materials. The Chattanooga Creek Site is the 7.5-mile segment between the Tennessee border and the Tennessee River. It is one of numerous creeks in Hamilton County that has suffered the severe consequences of industrial and municipal degradation. A large number of industries, including those engaged in the manufacture of coke, organic chemicals, wood preserving, metallurgical and foundry operations, tanning and leather products, textiles, brick making and pharmaceuticals discharged untreated industrial waste into the creek for over seventy years. The flood plain of the creek was also used by the City of Chattanooga, the industries and private citizens as a dumping ground of municipal and industrial refuse and waste. Many groups have worked to address the problem—the University of Tennessee at Chattanooga, neighborhood groups, the Agency for Toxic Substances and Disease Registry, the Environmental Protection Agency, Tennessee Valley Authority, the City of Chattanooga, and the Tennessee Department of Environment and Conservation. As is often the case with Superfund sites, there is significant disagreement about the best way to clean up the site, and how clean it needs to be. In Chattanooga, much effort is expended placing blame for slow remediation on "inflexible [government] regulations that exclude innovative technologies and quicker solutions" (Sustainable Chattanooga 1995, 1).

In addition to the Chattanooga Creek Cleanup effort, a number of other proactive projects, such as the North Chickamauga Creek watershed protection project, are underway. The Tennessee Valley Authority has established a Clean Water Initiative bringing together multidisciplinary River Action Teams (RATS) to work in partnership with local officials to prevent and clean up pollution in twelve major watersheds around Chattanooga in the Tennessee River system. This also includes a

Wellhead Protection Plan, which is a partnership effort between the Regional Planning Commission, Tennessee American Water Company, and eight utility districts. Overall, significant progress appears to have been made to protect Chattanooga's water resources.

One of the more comprehensive efforts to protect and improve water, particularly among efforts that contain explicit indicators and goals, is found in draft plan in San Francisco. This plan specifies a number of water-related goals and objectives both for the short-run (5 years) and the long-term. These goals include:

To establish reliable drinking water supply and quality. To establish consistent water pricing. To use wastewater for the greatest possible number of uses that can be shifted from Hetch Hetchy water. To assure that water delivered to San Francisco is safe from potential contaminants such as herbicides and pesticides. It should be of sufficient quality that it will not corrode water pipes, which could put excessive amounts of copper, lead or other metals into the water. To limit the chemicals used in water treatment and storage to only what is necessary to meet safe drinking water standards. To establish conservative estimates of available water supplies, accounting for prolonged periods of drought. To include in estimates of available water supplies the need to support all water users and maintain regional biodiversity. To protect urban watersheds from development that could have an impact on the resource. To protect groundwater from contamination and salt-water intrusion. To monitor and control withdrawals of groundwater to assure groundwater availability and quality for periods of drought and create an adequate emergency supply. To reduce water consumption and maximize the use of recycled water. To maintain water storage and conveyance structures in good repair. To encourage wastewater recycling. To assist in improving the quality of the San Francisco Bay and the Pacific Ocean water discharges should be cleaner than the background quality of these water bodies. To reduce liquid wastes at the source rather than through the chemical treatment processes. To eliminate toxic chemicals at water pollution control plants. To ensure that water pollution does not occur from discharges from the solid waste stream (i.e., the garbage). To reduce the adverse impacts of storm-water runoff into San Francisco Bay and the Pacific Ocean. (City of San Francisco 1995, 24–25)

These goals are accompanied by a number of specific indicators, including:

Reused wastewater consumption per capita; percentage of wastewater reused; percentage of public landscaping in drought-resistant plantings; the volume of stored water; percentage of public schools including water management issues in curriculum; the volume of water discharged into San Francisco Bay and ocean; the number of times out of compliance with State potable water quality standards; and the number of times out of compliance with State discharged water quality standards. (City of San Francisco 1995, 24–25)

The short-term objectives to be achieved by the end of 2001 for these goals and indicators called for the city to:

increase reuse of treated wastewater by 10%; increase the number of dual-plumbing systems by 5% per year over the preceding five years; increase reuse and/or sales of reclaimed wastewater byproducts by 10% a year; establish a conservation and reclamation demonstration project in a medium-density residential apartment building, an industrial park, a major school facility and a medium downtown commercial building; establish one potable water reuse demonstration project; and perform commercial water conservation audits for 20% of San Francisco businesses. (City of San Francisco 1995, 24–25)

Clearly, San Francisco's initiative places a great deal of emphasis on finding ways to recycle and reuse water resources as a way of conserving water from other sources. Although many cities seek to systematically manage their water resources, San Francisco's plan contains a level of details and specificity not often found in local sustainability initiatives.

Solid and Hazardous Waste Efforts

Managing the streams of solid and hazardous waste in cities constitutes an enormous challenge. Solid waste management, a traditional function of many city governments, has become more difficult and more costly as greater restrictions have been placed on landfills and incineration. As difficult a challenge as solid waste management poses, hazardous waste issues are even more difficult. Whether in the form of efforts to clean up existing hazardous waste sites, including National Priorities List, or Superfund, sites, or of efforts to monitor and manage streams of hazardous materials and wastes, cities face a daunting task. Virtually all city efforts at solid waste management include a healthy dose of recycling. Cities vary in the extent of coverage of their recycling programs, sometimes limiting the types of materials that are included in the programs. Cities vary in the extent of residential participation in recycling, and in the proportion of the waste stream that is recycled. It is increasingly common for cities to expand their recycling programs to include household hazardous waste, although many cities still do not have such programs. Sustainable cities often tend to be more ambitious in their solid and hazardous waste management, as examples from Jacksonville, Austin, and San Francisco demonstrate.

Reducing the Amount of Solid Waste and Recycling Solid Waste   There are a number of approaches to solid waste and recycling pursued in sustainable cities. Household solid waste recycling programs exist in virtually every city in the United States. Cities that take sustainability seriously usually make concerted efforts at recycling and reuse of solid waste. Sustainable indicators projects frequently elect to monitor one or both of two measures of household recycling: a measure of the household participation in recycling (which is usually measured by the proportion of households placing their recycle bins out on the designated days), or a measure of the amount of the total solid waste stream that is recycled (per capita, or as a proportion of the total waste stream). Additionally, cities often set specific targets for increased recycling or reduced landfill or incinerated solid waste.

Jacksonville's Quality of Life Indicators program specifies an explicit goal of reducing annual per capita tons of solid waste deposited in city landfills to 0.74 tons. Mainly through extensive recycling efforts, the city has decreased its per capita solid waste from 1.63 tons per person in 1987 to 0.82 tons per person in 1996, although that number increased to 0.96 in 1998. Jacksonville's indicators do not specify particular recycling goals, but the indicators project also includes a measure of recycling. The city has used, since 1989, an indicator of the per capita tonage of solid waste processed for recycling as part of its city-wide residential recycling program instituted in 1990. This indicator shows that household solid waste recycling increased considerably from .41 tons per person in 1989 to 1.01 tons per person in 1995, but has since dropped back to .74 tons per person in 1999. The city has set no stated targets for the future.

This is similar to the indicator used in the Central Texas Indicators 2000 program, which includes Austin (the Central Texas indicator focuses on the total weight, opts to measure the weight of landfilled solid waste in pounds rather than tons). This program does not specify a particular goal, but the indicator suggests that the Austin area disposed of about 3,139 pounds (1.57 tons) of solid waste in 1998, up from the 2,920 pounds (1.46 tons) of waste landfilled in 1996.

The San Francisco Sustainable City initiative has sought to reduce the amount of household solid waste produced from 7.5 pounds per person per day to 6.0 pounds. San Francisco's Sustainable City initiative articulated a number of specific goals for the 2001 year, including recycling 50% of the solid waste stream by diversifying what materials are recycled, getting 20% of the city's businesses to commit to buying recycled paper and other recycled products, expanding backyard composting to 15% of the city's residences, and training at least 1% of the city's residents in source reduction practices.

Reducing the Amount of Hazardous Waste   For the most part, handling of industrial hazardous wastes within a city's borders is regulated by federal and state laws, and city governments do very little beyond what these laws require. Some cities, however, do sometimes build measures of the amount of hazardous waste produced or disposed of in the city with associated efforts to reduce these, and they often make efforts to try to ensure that industrial, and otherwise unregulated household, hazardous waste is managed in particular ways.

Industrial Pollution Remediation, Prevention, and Reduction Efforts
Perhaps one of the environmental areas that most distinguishes cities that take sustainability seriously is whether they engage in explicit forms of pollution prevention, reduction, and remediation activities. Most cities do very little in these areas, and even cities that have fairly well developed sustainability plans often seem weak on pollution prevention. Yet a few cities have been very aggressive in trying to find ways of intervening to reduce or prevent pollution beyond that which is accomplished by virtue of state or federal laws.

Pollution Reduction   Few cities have pursued specific policies or programs to promote anything close to comprehensive pollution reduction or pollution prevention. Perhaps one of the more impressive efforts to do so comes from Boulder. Boulder operates a collaborative program of pollution prevention called Partners for a Clean Environment, or PACE. PACE's effort to reduce air pollution through reductions of volatile organic chemicals was discussed earlier and includes initiatives to reduce the volume of hazardous materials from small businesses. For example, it includes an effort to minimize disposal of petroleum-based solvents, antifreeze, and used oil filters from automobile repair shops. It also includes an effort to reduce toxic material disposal from print shops (PACE 2000).

Eco-Industrial Parks   Many cities have elected to focus their economic development and growth efforts on the creation of eco-industrial parks. As part of an overall plan of "smart growth," discussed more fully elsewhere in this book, eco-industrial parks represent an effort to allow or encourage economic growth while minimizing environmental impacts. An eco-industrial park is a concentrated area of industrial activity much like a standard industrial park, but with a difference. In some places, an eco-industrial park might simply consist of a concentration of businesses selected because they, by the nature of what they do, produce minimal environmental damage or threat. Yet the concept goes well beyond this. The more promising idea is that an eco-industrial park is a collection of industries that are selected and located in close proximity to each other because of their particular relationship to one another so that one company's waste would be another company's factor of production. As stated in the Greenlight Foundation's 1999 Earthday press release, an eco-industrial park is where:

one industry produces by-products that are then converted into new products by other park tenants. Additionally, park tenants are linked into a supportive community through codes and covenants that help them respond to community values. For those communities that work to build eco-industrial parks, opportunities for sustainable economic health are promised. Sustainability is promised because successful tenants of the park will have reduced incentives to relocate. Tenants will secure a package of benefits customized for their needs that will be quite difficult to capture elsewhere. Opportunities for community health can occur when the relationship with the local ecological system is identified, thus reducing the load on resources through more effective or efficient use. Eco-industrial development can create jobs and a healthier environment at the same time. (Greenlights Foundation 1999)

There are many variations of eco-industrial parks and many configurations or collections of types of industries. Perhaps one of the most notable examples of such a park comes from Chattanooga, where the city has combined the goal of creating manufacturing jobs with an effort to be more environmentally responsible. In Chattanooga, Pittsburgh, Chicago, Minneapolis, and Brownsville, Texas, to name a few places, the idea of developing eco-industrial parks has received significant attention in the last half decade.

Superfund Site Remediation and Brownfield Redevelopment How do sustainable cities deal with the presence of hazardous waste sites within their borders? Virtually every major city has some degree of problem with hazardous waste sites, including but not limited to those listed on the National Priorities List (Superfund) under the federal Comprehensive Emergency Response, Compensation, and Liability Act of 1980. For the most part, these issues are confronted in the context of what are often referred to as "brownfields." A product of modern industrialization, urban brownfields consist of land that has been rendered unusable by the existence of, or the perceived existence of, environment contamination. The U.S. Environmental Protection Agency's working definition of a brownfield is "abandoned, idled, or under-used industrial and commercial facilities where expansion or redevelopment is complicated by real or perceived environmental contamination" (USEPA 2002). For all practical purposes, brownfields consist of tracts of land in urban areas that sit idle because the cost of converting the land into any sort of productive use exceeds that which any market would bear. Often, the existence of contamination ensures that no private investor will capitalize redevelopment because of the associated financial and legal liabilities that would be incurred. Rather than simply allowing such land to be unproductive or detrimental to the livability of cities, city governments have increasingly engaged in some form of brownfields redevelopment. Stimulated by an extensive pilot program by the EPA (USEPA 1998), cities all around the country have made significant efforts to convert brownfields into economically and socially productive resources.

Virtually every major city in the United States has some amount of land that has been designated as a brownfield or that could be given that designation. Most of these cities have engaged in some sort of initiative to redevelop at least some of the brownfield land (Russ 2000). However, most cities have not placed this redevelopment explicitly into the context of sustainability per se. In most cities, brownfield redevelopment is treated as just another form of economic development where the land is developed according to its "highest and best use" given the economic constraints it faces. Using brownfields redevelopment as an opportunity to pursue sustainability is often given less import.

Starting in 1998, the EPA began making a special effort to encourage cities to place brownfields initiatives into such a sustainability context, and has produced the conceptual frameworks to guide how this can be accomplished. These frameworks stipulate the multiple goals that need to be achieved to ensure that brownfield redevelopment is consistent with some conception of sustainability, and lay out the characteristics that such redevelopment would possess (USEPA 1998; USEPA 1999). In broad-brush, these guidelines suggest that as communities decide what to do with the brownfields land, they must consider how the proposed use will contribute to ecological, economic, and social integrity. The frameworks are very much process-oriented, seeking to put in place community processes that are designed to maximize the chance that the resulting redevelopment projects and activities will contribute to sustainability. Although there have been many examples of sustainable brownfields projects, there are three cities, Baltimore, Portland, Oregon, and Chattanooga, whose initiatives provide impressive results within the context of sustainability. All three cities have been selected as EPA pilot brownfields project cities.

Baltimore   The City of Baltimore estimated that 3,500 to 5,300 acres of land zoned for heavy manufacturing contains environmental problems that impair their marketability (Baltimore Brownfields Project 2000). In particular, the City identified several sites located in Baltimore's Empowerment Zone where contamination presented a substantial obstacle to economic revitalization. Baltimore's brownfields initiative sought to balance economic growth and redevelopment while providing appropriate and sufficient protection of the environment, especially the Chesapeake Bay watershed area. Brownfields redevelopment in the City has sought to promote efficient land-use patterns, while reducing the air and water pollution associated with urban sprawl, and expanding job opportunities in locations that are accessible to lower-income populations. The pilot project identified and targeted several prospective brownfield sites within the Empowerment Zone, and has used this as an opportunity to pursue the development of an ecological industrial park. The initiative has developed a partnership development between Baltimore City departments and the Baltimore Development Corporation to develop a comprehensive inventory of vacant and underutilized industrial properties in the City. When fully developed, this inventory is expected to serve as an economic redevelopment tool and planning tool, and allow users to query for sites with desired specifications, and search for strategic land assembly opportunities.

The specific activities of the Baltimore brownfields pilot project include facilitating the process for assessment, remediation, and redevelopment of the former 33-acre ASARCO smelting site. There have been 200 short-term remediation and construction workers employed at the site demolishing parts of and renovating other sections of the 750,000 square-foot complex. When complete, the port-related waterfront warehouse space is expected to provide full-time employment for 180 employees. Over $11.5 million was invested from both public and private sources to enable this project to move forward. Based on referrals from the Brownfields Industrial Redevelopment Council, two sites were committed for redevelopment into a 115-unit townhouse project and a drug store. Partnering by the city government and the Empower Baltimore Management Corporation, the official federal urban empowerment zone, has resulted in the creation of a $3 million revolving loan and grant program dedicated to brownfields revitalization projects in the Empowerment Zone. The City is pursuing remediation and redevelopment at two additional sites with employment potential of at least 50 jobs.

Portland   In February 1996, the U.S. EPA selected the City of Portland for a national brownfields pilot (Portland Brownfields Initiative 2000). As the state's oldest and largest industrial, shipping, and commercial center, Portland has a high concentration of abandoned or underutilized properties surrounding the City's Enterprise Community where unemployment is 10.4% and the poverty rate is 35%. The threat of contamination and liability has inhibited reuse and redevelopment at these sites while suburban sprawl continues. The ultimate goal of Portland's brownfields effort was to encourage economic growth and redevelopment at specific sites within its Enterprise Community and along the city's waterfront while protecting the environment, especially the Willamette River. Redevelopment of the urban waterfront promotes productive land use, reduces air and water pollution associated with urban sprawl, and expands job opportunities in locations accessible to low-income populations.

Specific activities completed as part of this pilot program included conducting education and information exchange on brownfields issues to inform and involve citizens, creating outreach opportunities for schools and civic associations to raise awareness and promote development, and promoting redevelopment of specific sites through an Internet-accessible online computer information system that provides data on site assessments, cleanup, and development. All of this is intended to achieve coordination with interested stakeholders in order to develop focused action plans, and to focus the city's brownfields efforts and resources on specific identified sites within the Enterprise Community.

Chattanooga   The city's brownfields initiative has focused mainly in one specific area of the city. Approximately 5,300 people live in the project area of Alton Park, which contains at least 34 specific brownfield sites, of which 5 have been targeted by the pilot project. This targeted area has a poverty rate of 61%, with a median household income of $12,300. The area's population is 98% African American. The 2.7-square-mile Alton Park area was once home to the foundries, tanneries, brick kilns, glass container manufacturers, and textile mills that made the city a manufacturing center of the mid-Atlantic states. For a century, the economy in Alton Park was strong and the area bustled with industrial jobs, local businesses, and residential neighborhoods. But by the 1960s, new manufacturing methods and the changing economy led to the deterioration of the major industries in Chattanooga. Residents who could afford to moved to the suburbs, retail businesses closed, schools shut down, and there was a perception that many of the decaying buildings in Alton Park were contaminated. Chattanooga designated this area as a brownfield, and made it the focus of its coordinated initiative.

The Chattanooga/Hamilton County Brownfields Program claims to serve as a catalyst for business expansion and relocation, job creation, environmental cleanup, and neighborhood rebuilding. The main focus of the project has been to identify and assess contamination at the five targeted sites in the project area, as well as assist in the development and implementation of innovative ways to finance cleanup. Other activities include creating a brownfields inventory, establishing general site redevelopment criteria through the use of a community and stakeholder participation process, and creating cleanup and redevelopment plans that identify innovative financing mechanisms for cleanup and address ownership procedures.

Biodiversity

Most cities that claim to pursue sustainability do not explicitly take into consideration issues of biodiversity. Yet biodiversity issues are very clearly part of the conceptual foundations of sustainability. Biodiversity simply refers to the range of biological organisms living on the planet or in a given ecosystem at a given time. Because of ecological degradation, the earth has become less biologically diverse over time. Concern about biodiversity comes largely from the idea that greater diversity supports healthier ecosystems. As Ophuls and Boyan state:

Several practical reasons exist for preserving . . . biodiversity. . . . Wildlife performs free environmental services. Plants take carbon dioxide from the air and produce oxygen. They counteract greenhouse emissions. They regulate stream flows and groundwater levels, recycle soil nutrients, and cleanse pollutants from surface water. . . . Wild species are the original source of over 40% of the chemicals used in prescription drugs. . . . Wild species provide humans with needed products and services. . . . When a species is lost, it is not renewable. Its particular function or potential—its unique combination of genes—is gone forever. A given loss may or may not have direct consequences for people. . . . [but] it would be foolish to assume that the whole phenomenon is inconsequential. (Ophuls and Boyan 1992, 133)

The sustainability challenge, then, is to find ways of protecting the existing range of plants and animals that exist in a city's ecosystems or within its borders. Although a number of cities, such as Seattle, Washington, and Cambridge, Massachusetts, have tried to incorporate the pursuit of biodiversity into their sustainability initiatives, perhaps the best example of a city that has tried to pay close explicit attention to this is San Francisco.

As part of San Francisco's Sustainability Plan, an effort was made to incorporate a variety of indicators of biodiversity (City of San Francisco 1996a). So primary among the goals and objectives are efforts to achieve greater understanding of biodiversity (an education function), the protection and restoration of remnant natural ecosystems, and to protect species of plants and animals that are particularly sensitive, among others. Work on achieving these goals has been accomplished through city-wide educational and public information programs, local government purchase of green spaces, enforcement of state and local environmental regulations, the development of integrated pest management, and other activities.

Energy Use and Conservation

Closely linked to issues of the environment per se are issues of energy usage and consumption. Most of the air pollution visited upon America's cities comes from the burning of fossil fuels, whether it is gasoline in the automobile, fuel oil for heating homes, producing electricity, or other industries, the sources of energy and the amount of that energy consumed play an important part in influences the amount of air pollution emitted in a city. Sustainable cities frequently attempt to address energy issues by influence the consumption of energy, usually offering consumers public transportation alternatives to the single-passenger automobile for commuting to work, and home energy conservation opportunities, such as increased home insulation, and more efficient heaters and appliances. Some cities are able to offer consumers the option of receiving electricity generated from renewal sources, such as windmills, although in the wake of the nationwide frenzy to deregulate electric utilities, these efforts have apparently stalled, particularly in California. These measures and many others often characterize cities' sustainability efforts with respect to energy.

Santa Monica's Sustainable City program developed one central indicator of energy consumption, the number of mBTUs of electricity and natural gas used city-wide by all users, and is also broken down by the type of user, residential, commercial, and industrial. The original goal was to decrease the consumption of electricity and gas by 16% from its 1990 baseline, although this target was later dropped for technical reasons. Over the period from 1990 through 1997, energy consumption dropped from 6.45 million BTUs (mBTUs) to 5.77 mBTUs, although electricity use increased. All of the fluctuation in energy use was caused by variation in the industrial sector.

Santa Monica's energy plan calls for a variety of programs and activities by city agencies to increase energy efficiency and otherwise reduce consumption of nonrenewable energy sources. For example, starting in 1995, the city embarked on an ambitious program to replace heating, lighting, and cooling systems and their controls in all city buildings. In 1999, it also entered into a contract to purchase 100% renewable-source electricity (in this case, geothermal electricity) used by city government. And during 1999, the city developed projects to install photovoltaic electric generation capabilities in a number of city-owned facilities, including the Santa Monica Pier and the Civic Center. However, in the most recent Progress Report, the city expressed some reservations about whether the city would be able to achieve greater reductions in energy consumption through its programs because of the 1998 deregulation of the electric industry in California.

Another example of a sustainable city that makes a concerted effort to affect energy consumption is San Francisco. As part of its indicators project, adopted in 1997, San Francisco targets "energy, climate change, and ozone depletion." Its energy indicators focus on city-wide energy consumption, consumption of energy from renewable sources, fossil fuel sales, vehicle electricity and hydrogen use, the number of alternative fuel vehicles registered in the city, and several other indicators. The primary goal of increasing renewable relative to nonrenewable energy consumption has been pursued through a number of different programs and projects, including a program to pursue resource-efficient city buildings, education and outreach initiatives, and several dozen other narrowly targeted efforts (City of San Francisco 1996b).

One of the approaches to energy conservation in many sustainable cities initiatives is found in "green building" programs. Green building programs may provide building contractors and designers with specific information about how to design and build more energy efficient buildings. In some places, they provide direct incentives for incorporating energy efficiency into the design and construction of new and renovated buildings. Austin, Texas, has a green building program, started in 1993, that is limited to providing for a fee information to architects, builders, and homeowners, while making available a system for ranking new construction in terms of how "green" it is, where energy efficiency represents a significant element of the assessment (City of Austin 2000). San Jose started its green building program in 1998 (City of San Jose 2000).

In perhaps the most aggressive effort of this type, some cities have enacted local building code ordinances that require achievement of energy efficiency in new construction. In addition to the energy conservation efforts that apply to the operation of city government in Santa Monica, the city has also adopted a stringent "green buildings" program. This program consists of a series of municipal code ordinances that require new buildings to meet high standards of energy consumption. Indeed, the most recent ordinances require building design features to produce 20 to 25% more efficiency than California's fairly strict statewide regulations. The city ordinances specify the required efficiency of windows and doors, appliances, lighting, and so on. In order to obtain building permits, building designs must comply with these ordinances.

Boulder, Colorado, makes demonstration of energy efficiency an explicit part of the process of getting building permits. Boulder's "Green Points" initiative requires that applications for building permits on new residential construction must earn at least 25 green points before they will be granted. Green points are earned according to the amount of insulation (for example, R-24 wall insulation would earn 3 points), the type of windows (high performance glazing would earn between 2 and 8 points), the kind of heating system (radiant floor heating, for example, would earn 5 points), and by positioning new construction so that it has enhanced access to solar heat (2 points) (City of Boulder 2000). Once a building permit has been issued, the design features that earned the green points are subject to building inspection just as with any other elements of the building code.

One of the more innovative programs to alter the source of household electricity is found in Austin. Austin owns the city's electric generating company, called Austin Energy, Inc. It has embarked on an effort to offer consumers clean energy alternatives, spurred by the city council's resolution that 5% of the city's electricity should come from renewable sources by 2005. The utility has created the "GreenChoice" option for consumers, where they can opt to be supplied with electricity generated from renewable sources (wind, solar, and biogas). Renewable source electricity is priced slightly higher than fossil fuel generated electricity, but that price is guaranteed for up to ten years, while electricity generated with coal, fuel oil, or natural gas carries no such guarantee and is subject to the vagaries of the fuel markets. To ensure sufficient supplies of renewable electricity, the utility operates a wind turbine farm in West Texas, with significant expansion planned in the years to come.

Public Transit and Transportation Planning

Energy usage and energy conservation are closely linked to issues of transportation. Indeed, the existing literature on sustainable cities suggests that the single greatest threat to achieving sustainable results in the United States comes from America's addiction to the automobile. The argument concerning the way that reliance on automobiles for transportation, particularly to and from work, undermines sustainability couldn't be clearer. As Newman and Kenworthy state,

. . . the problem of the car in cities is that the freedom and power it gives us come at a cost. It is easy to see some of the case-based environmental costs in polluted air, noisy environments, and acres of asphalt for parking and roads. But some problems such as urban sprawl . . . are also fundamentally due to an overemphasis on cars that facilitates dispersed, low-density suburbs. Even stormwater pollution is found to be greater in car-based cities due to the higher amount of hard surface. (Newman and Kenworthy 1999)

Indeed, environmentalists have for years pointed to America's dependence on the automobile as a primary cause of environmental pollution, and as a problem that will not readily change for the better (Hart and Spivak 1993).

Although automobile use remains one of the more difficult sustainability issues to tackle, sustainable cities often engage in various kinds of transportation planning, and do so with an eye toward trying to give residents and commuters the opportunity to use public transit. Cities that take sustainability seriously try to integrate transportation planning with other types of planning, including residential planning and zoning, industrial and job site location, and other issues. A number of cities' sustainable indicators projects incorporate transportation issues. Additionally, many cities' Comprehensive Planning is used to coordinate transportation in a metropolitan-wide fashion, and to define their urban growth boundaries in such a way as to minimize reliance on the automobile, particularly for commuting to work (Dunphy et al. 1996; Newman and Kenworthy 1999). This issue is explored more fully in chapter 4 as a form of "smart growth."

There are surprisingly few good examples in the United States of robust city or metropolitan efforts to affect reliance on the personal automobile. There are many prescriptions for how cities could begin reinventing their transportation planning, and some of the best examples of actual results in North America come from Canada (Newman and Kenworthy 1999, 212–223). Several of the cities on the sustainable cities list developed here, Boulder, Portland, and Boston, have been identified as U.S. cities that show "signs of hope" (Newman and Kenworthy 1999, 223–233).

Perhaps one of the more exciting efforts to link pollution reduction and transportation planning comes from Chattanooga. Advanced Vehicle Systems, Inc. (AVS) was organized in late 1992 in response to the Chattanooga Area Regional Transit Authority's (CARTA) decision to request bids for a manufacturer to develop and produce battery powered electric buses. AVS was started with the express purpose of producing the electric buses needed in the city and metropolitan area, and was awarded an initial contract to produce twelve electric buses for a three-mile shuttle system in a renaissance area of downtown Chattanooga. Subsequently, the electric bus initiative was enhanced by the creation of "The Living Laboratory," a collaborative effort involving AVS, the Electric Transit Vehicle Institute, the Tennessee Valley Authority, and CARTA. The idea was to marshal a variety of resources centered on electric and hybrid electric vehicles in one industrial location.

Summary

Cities that take sustainability seriously typically engage in a wide variety of activities that attempt to directly improve or protect the environment, and that try to do this indirectly through influencing energy consumption. Many cities attempt to go beyond what might be prescribed in federal or state laws to work toward reducing air emissions, improving and protecting sources of ground and surface water for drinking and recreational uses. Nearly all cities engage in efforts to reduce their streams of solid waste, although sustainable cities are often more aggressive about such efforts, especially in expanding household recycling programs. Cities sometimes attempt to affect the way hazardous wastes are managed, and the ways that hazardous waste sites are remediated. As part of the remediation process, some cities pursue policies to redevelop brownfields sites. And a few cities make explicit efforts to protect biodiversity. Increasingly, cities that take sustainability serious rely on clusters of related economic development in eco-industrial parks so they can minimize the ecological footprint imposed by the industrial sector of the local economy.

In one way or another, whether in the context of resource conservation, or pollution reduction, or public transportation policy, or other more general goals, sustainable cities often seek to reduce reliance on nonrenewable energy sources, particularly fossil fuels. Many cities have built energy consumption issues into their sustainability indicators, usually focusing on measuring residential, commercial, and industrial energy usage. Some cities have access to electricity produced by windmills or through geothermal generation, and they seek to substitute electricity from these sources for power generated from burning fossil fuels or from nuclear generating stations. Cities regularly embark on public education and outreach campaigns to convince people to change their energy consumption. And cities that seem to take sustainability seriously have made significant changes to their local building codes to require greater energy efficiency in new construction and in renovation.

4

The Economic Development Side of Sustainability: Growth versus Smart Growth

All cities and towns feel the need to pursue economic growth and engage in economic development. Nearly all adopt policies and programs designed to affect economic development and make efforts to ensure that there is an adequate base of employment for their residents. Every city and town understands that there is some relationship between the function of the local economy and the quality of life that residents are able to enjoy, but sustainable cities tend to view economic development differently from other cities. Sustainable cities tend to see development as a means to an end, a means to achieving a particular type and level of quality of life, locally defined but potentially including a wide variety of quality of life characteristics. Unlike many cities, where economic growth is the imperative for its own sake, or where quality of life is itself defined just in terms of the quality of employment and average family incomes, or where there is merely an assumed relationship between economic growth and quality of life (i.e., more growth means a better quality of life), sustainable cities seek to manage economic growth and development to be more consistent with their visions of what kind of community they desire to achieve. For these cities, gone are the days when leaders were willing to assume that more economic development is necessarily always better. Here are the days when leaders try to pick and choose economic development that directly serves what they perceive to be the community's vision.

Not all cities have the luxury of being able to choose what kinds of economic development they will accept, and indeed the "beggars can't be choosers" mentality is alive and well in many U.S. cities. The mindset that many city governments bring to economic development is that "we need them more than they need us." As a consequence, cities often knowingly accept economic development that provides less net benefit than they might like. Cities that are able to be more selective about economic development often can do so because of some natural or cultivated comparative advantage. Geographic location and proximity to natural resources, availability of inexpensive land, access to a skilled labor force, and statewide and regional public policies all contribute to providing some cities with advantages they can turn to their benefit when it comes to defining and pursuing their brand of economic development. However, care must be taken not to read too much into the fact that some cities are able to be selective about the kinds of economic development they will accept. Some cities exist in a metropolitan area or region where they are essentially subsidized by the economies of neighboring places, able to avoid some kinds of less desirable economic development within their borders by virtue of having the benefits of economic development that occurs outside of their borders. Of course, in environmental terms, this would necessarily mean that the city's ecological footprint is far more expansive than a city with no such advantages.

This chapter examines some of the ways that sustainable cities have tried to manage their economic growth and development. Before examining these ways, it is necessary to build a picture of how economic growth and development are often seen and pursued in cities that are not particularly concerned with sustainability. When local leaders and residents see city government as a tool of "the city as growth machine," the pursuit of sustainability is virtually impossible. When city leaders perceive that there is a single imperative for local policies, and that imperative is to maximize economic growth, then the idea of sustainability becomes almost an anathema. Perhaps the most intriguing aspect of this is the fact that many cities one might expect to be driven by "growth machine" conceptions of economic development—cities that were once dominated by such conceptions—now have adopted some form of "smart growth" approach.

Traditional Conceptions of Economic Growth and Development in Cities

Although there are many different ways of viewing the role of economic growth in American cities, one concept focuses on the absolute imperative of the pursuit of economic growth and development to city policymakers. Under this conception, all other policy and program issues take a back seat to the need to create growth and its corollaries. As noted earlier, there is often an assumption dominant in city politics that economic growth is good, period. This view is often supported by a local coalition of public officials, influential business leaders, real estate developers, labor leaders, and others, to push the idea that more development is simply better. As stated by Judd and Swanstrom,

> The members of the downtown growth coalitions are convinced that the city's survival depends upon local economic vitality. The logic of their argument is as follows: A healthy tax base is necessary to maintain the public services and infrastructure that the local economy and residents rely upon; a healthy tax base, in turn, requires rising land values and business prosperity; in order to attract private investment, a city must offer a good business climate and special incentives that make it less expensive for mobile investors to locate there than in other cities; an increasing volume of private investment begins the cycle all over again, resulting in rising land values, a healthy tax base, better services, and a better business climate. (Judd and Swanstrom 1998, 359)

Perhaps the strongest proponent of this view is Harvey Molotch, who describes this imperative with reference to the city as a "growth machine" (Molotch 1976). According to this argument, the traditional city is dominated by attention to growth. As he states it:

> ... the political and economic essence of virtually any given locality, in the present American context, is *growth*. ... The desire for growth provides the key operative motivation toward consensus for members of politically mobilized elites, however split they might be on other issues, and that a common interest in growth is the overriding commonality among important people in a given locale. ... This growth imperative is the most important constraint upon available options for local initiatives in social and economic reform. ... The very essence of a locality is its operation as a growth machine. (Molotch 1976, 309–310)

Thus, in traditional conceptions, growth is seen as the engine that drives the health of the city. Cities that experience growth are healthy, and that health has definite characteristics. Molotch describes these characteristics when he notes that:

> the clearest indication of success at growth is a constantly rising urban-area population—a symptom of a pattern ordinarily comprising an initial expansion of basic industries followed by an expanded labor force, a rising scale of retail and wholesale commerce, more far-flung and increasingly intensive land development, higher population density, and increased levels of financial activity. (Molotch 1976, 310)

Cities that do not have these characteristics, those whose economies are stagnant or in decline, are not healthy according to this view. Yet embedded in this traditional conception of growth is an assumption that these characteristics and corollaries of growth are good, that they produce better quality of life results for residents. As noted earlier, this assumption is called into question by advocates of sustainable communities and of smart growth alternatives (Young 1995). The purpose of this discussion is not to refute the argument that economic growth is important to the health and vitality of cities. The purpose is to contrast this growth machine view of economic growth with an alternative view that seems to disproportionately characterize the mind-set found in cities that seem to take sustainability seriously. Like Clarence Stone's (1993) "middle-class progressive" cities, that begin to address environmental protection, affordable housing, the quality of urban design, and the creation of linkage funds to support a variety of social purposes, many cities take a different view toward economic growth and development, and this view is largely captured by the idea of "smart growth."

Sustainable Conceptions of Economic Development: Smart Growth

Recent conceptions of economic development attempt to ensure that economic growth is explicitly tied to the quality of the life in the community. More accurately, such conceptions try to ensure that whatever economic growth and development takes place in the city is consistent with some vision of what the quality of life in that city should be. This conception, often referred to as "smart growth," expresses concern about unmanaged development and its consequences. It recognizes that unmanaged development produces a variety of consequences that in a particular community might not be desired. Urban sprawl, abandoned and deteriorating inner-city infrastructure, traffic congestion, declining public services, particularly in public education, and many other urban pathologies, as well as the ultimate waste of natural and human resources that result, have been laid at the feet of unmanaged or improperly managed economic development (About Smart Growth 2001). Additionally, over time, unmanaged or improperly managed regional and metropolitan economic development have ensured that these problems would spread from inner-cities to surrounding areas including suburbs.

The alternative, smart growth, designs aid economic development in an effort to avoid creating these problems. As Geoff Anderson, of the International City/County Management Association, notes:

Smart growth recognizes connections between development and quality of life. It leverages new growth to improve the community. The features that distinguish smart growth in a community vary from place to place. In general, smart growth invests time, attention, and resources in restoring community and vitality to center cities and older suburbs. New smart growth is more town-centered, is transit and pedestrian oriented, and has a greater mix of housing, commercial and retail uses. It also preserves open space and many other environmental amenities. But there is no "one-size-fits-all" solution. Successful communities do tend to have one thing in common—a vision of where they want to go and of what things they value in their community—and their plans for development reflect these values. (Anderson 1998, 4)

The connection between smart growth and sustainability seems perfectly clear to advocates and practitioners. As Goode, Collaton, and Bartsch suggest,

Smart growth takes aim at the programs and policies that literally drive development away from existing communities. Veterans of urban planning argue that the issues pushing the smart growth debate, while not new, have emerged into a fundamentally different approach to maintaining metropolitan areas as sustainable places to live and work. (Goode, Collaton, and Bartsch 2000, 1)

In a similar vein, Kline recognizes the connection between economic growth and the values that underlie visions of quality of life articulated in particular communities. For her, the key is what she calls "economic security," which is meant to convey the idea that some aspects of economic growth are directly linked to quality of life issues, although these aspects may not be those that are most commonly assumed to produce better quality. As Kline notes,

A more sustainable community includes a variety of businesses, industries, and institutions which are environmentally sound (in all aspects), financially viable, provide training, education and other forms of assistance to adjust to future needs, provide jobs and spend money within the community, and enable employees to have a voice in decisions that affect them. A more sustainable community also is one in which residents' money remains in the community. (Kline 1995, 4)

These two conceptions of economic growth and development, the view of the city as growth machine, and the view that growth ought to be pursued only in ways that foster particular types of social results, stand in sharp contrast to each other. The former reflects a view that economic development ought to be largely a matter of private investment. In this view, government ought not to "interfere" with these private investment decisions. As a consequence, residents of the city experience whatever quality of life results from the aggregate activities of the individual people in the city. For many advocates of unfettered economic growth, the issue is which approach maximizes economic growth, and which approach can be shown to be more "economically efficient." Indeed, studies of the economic efficiency of regulations on economic growth vis-a-vis the absence of such regulations do tend to support this argument (Charles 1998; Franciosi 1998; Peiser 1989), although other studies are less definitive (Breslaw 1990; Fischel 1990; Gordon and Richardson 1998). The latter view, that the economic efficiency of economic development is less important than the social effects of unfettered growth, reflects a view that unfettered private economic development decisions, such as the events that produce the tragedy of the commons described in chapter 5, sometimes undermine quality of life pursuits for the community as a whole. In practice, of course, cities often incorporate a mix of these two conceptions in their policies and programs.

Sometimes cities try to achieve some sort of balance in economic development, creating frameworks within which development can take place. Sometimes this balance includes a healthy dose of zoning and land use regulation. However, pursuit of smart growth often requires types and levels of regulation that are far more aggressive than that required for more traditional forms of development. Of course, not all cities actually possess the legal authority to regulate their economies or their land uses the way they might like. State laws often make it difficult or even impossible for cities to constrain development or to manage land uses. But within the range of their legal authority, sustainable cities tend to do more to manage economic growth and development than other cities. Indeed, it is not likely a coincidence that in states where some form of smart growth legislation or comprehensive planning requirements have been enacted, as in Arizona, California, Washington, and Florida, there are multiple cities pursuing sustainability. Yet many cities seem to be willing and able to adopt sustainable approaches to economic growth even in the absence of smart growth legislation in their respective states.

There has been a growing trend for impediments to smart growth to be redressed in cities and states through the referendum and initiative processes. Many ballot measures to control, regulate, or limit growth have been put forth over the last few years, and citizens groups have sponsored most of these. According to Myers and Puentes (2001), in the election of 2000, there were nearly 100 such ballot questions around the country, with at least 54 ballot questions dealing with land use and zoning authority, 15 on urban growth boundaries, and 2 comprehensive growth management plans. Five out of the 7 city ballot questions addressing growth management and regulation measures passed. Many of the other measures did not pass, but they do suggest increased attention to the need for greater growth management. City ballot questions providing funding for specific sustainability projects experienced mixed success. In Austin, a ballot question to create a light rail system was defeated, but in Seattle a new monorail system was passed. In Austin and San Francisco, initiatives to improve pedestrian safety and promote non-motorized transportation (bicycles) were approved. Proposals to protect undeveloped land from development in the form of open space were very successful. Of the 19 open space ballot questions that were voted on in U.S. cities, 18 passed.

Similarly, in the elections of 1998, ballot questions in many cities made an effort to affect local development. According to Myers (1999), communities are increasingly willing to finance "green infrastructure" by adopting proposals for open space acquisition. As she notes, ". . . Of the [226 local ballot measures in 1998], 163 measures were approved. . . . While the local measures represent a small proportion of the funding approved . . . they are a telling indicator of growing grassroots interest in conservation, outdoor recreation, and open space funding programs" (Myers 1999, 9). In Austin, Texas, for example, voters approved a park and greenway bond issue and a land acquisition bond issue that were part of the city's smart growth initiative. On the other hand, Portland, Oregon, defeated a city park bond issue, although another large open space bond issue was approved there a few years earlier.

Additionally, some ballot measures were used to roll back or limit the effectiveness of efforts by cities or states to control growth and sprawl. In Oregon, for example, a taxpayers group was successful in 2000 in its effort to pass a measure that requires the state or local government to fully compensate landowners for any reduced property values experienced as a result of land use regulations. In general, these ballot measures provide opportunities for a variety of sustainability related legal and finance issues to be addressed, sometimes enabling cities to pursue policies and programs that otherwise could not be accomplished. Regardless of whether an initiative is necessary, cities have taken a variety of approaches to smart growth. A brief review of some of the activities and approaches of sustainable cities toward smart growth will help to build a picture of how differently these cities pursue economic development.

Economic Development in Sustainable Cities

There is sometimes a tendency to equate concern for the environment with advocacy of no-growth economic development policies. Since economic development, and all the activities that are associated with economic development, is largely responsible for environmental degradation, so the argument goes, if a city wishes to be environmentally responsible, it must seek to limit or stop economic growth. Moreover, opponents of growth management often suggest that environmental concerns are used by antigrowth interests as a mechanism to block local growth efforts altogether (Frieden 1979). Yet this is far from the way that sustainable cities treat the issue of economic development. Cities that take sustainability seriously understand the role of economic development and economic growth in creating the kind of city they seek. As noted before, there is a clear recognition that controlled and balanced economic development is necessary to create and sustain the quality of life that residents seem to want. Exactly how much economic development and what kinds of economic development are, of course, the central issues cities must confront. A brief review of some of the issues and approaches to economic development practiced in sustainable cities paint a picture of how this implicit balancing is accomplished.

From a political perspective, smart growth provides an approach to economic development that carries many advantages compared to approaches that use other conceptions or terminology. Cities that have engaged in sustainability initiatives often discover that they have a difficult time connecting with the business community, and that getting business leaders involved in cooperative efforts is an enormous challenge. To business leaders, sustainability is a code word for being an environmentalist, and, with some notable exceptions, few self-respecting local business leaders wish to be associated with environmentalists. So, in many cities, pursuing sustainability is alienating to the business community, and any programs, projects, or specific initiatives that need business cooperation are unlikely to get it. The language of smart growth, on the other hand, is less likely to be seen as carrying the environmentalist connotations, and business leaders may well feel more comfortable being engaged in smart growth activities and planning processes. After all, who could be against smart growth if the alternative is dumb growth?

The backdrop that plays a central role in determining how far a city can go in managing its economic development, regardless of what form that development takes, is state law. As creatures of the states, cities are not always free to manage economic growth the way they might like. Some states provide the statutory foundation for city, county, or regional growth planning, and others do not (Smart Growth Network 2001). State law in Georgia, for example, makes it very difficult for cities such as Atlanta to intervene and control the kinds of economic development taking place there. And among those states that do provide some statutory foundation for growth planning through the creation of urban growth boundaries, such as California, Maryland, Oregon, Washington, and Tennessee, the motivations can be quite varied (English, Peretz, and Mandershield 1999; Nelson and Moore 1996; Washington State 1990). In Oregon, for example, one of the key reasons for state-mandated growth planning is to provide the legal foundation for cities in the state to place locally responsive constraints on growth. Cities in Oregon can designate how and where economic growth will take place when they designate their urban growth boundaries (Nelson and Moore 1996). In Tennessee, where growth planning also requires cities to designate areas where growth will take place, one key reason is to ensure that cities will not be able to legally prevent growth. With this backdrop, however, cities are able to engage in a variety of economic development activities that might be said to be consistent with smart growth approaches.

But what kinds of economic development activities do sustainable cities engage in? What kinds of processes are used to make these activities happen? And in what ways might these activities be said to constitute elements of taking sustainability seriously? Cities' smart growth activities includes such programs as brownfield redevelopment, as discussed in chapter 3, the use of land use planning and zoning, targeted or cluster economic development, eco-industrial parks, and the creation of ecovillages, to name a few prominent examples. A brief overview of the smart growth activities of several cities should begin to demonstrate this approach.

One of the better examples of basic smart growth economic development comes from the cluster development pursued in the city of Chattanooga, and involves the development of the local capacity to manufacture electric and hybrid vehicles, as discussed earlier. This story begins in 1992, when the Chattanooga Area Regional Transit Authority (CARTA) decided to request bids for a manufacturer to develop and produce battery powered electric buses for use mainly in Chattanooga. Advanced Vehicle Systems, Inc. (AVS) was organized in late 1992 in response. AVS was successful in being awarded the contract to produce twelve electric buses for a three-mile shuttle system in a renaissance area of downtown Chattanooga, and has since enlarged its line of products to include hybrid vehicles, and expanded its production and sales to other cities. This represents economic development that accomplishes two sustainability goals: it creates local jobs while at the same time contributing to improving the environmental health of the city. Chattanooga has been able to build on the successes of AVS by starting the Electric Transit Vehicle Institute (ETVI), a nonprofit organization formed to promote the design, production, and utilization of battery-powered electric and hybrid-electric vehicles. Subsequently, the Chattanooga Chamber of Commerce defined an economic development "cluster" in the area of electric vehicles as part of its regional development initiative.

Another example of a smart growth strategy employed in pursuit of sustainable development comes from Austin, Texas, where instead of reliance on regulatory (zoning) approaches, as in Portland, Seattle, and many other cities (as discussed later), the city creates financial market incentives to encourage economic development that is considered consistent with the city's overall vision for sustainability. Under this plan, a panel of city officials and developer's representatives assesses development proposals. This assessment assigns points to development proposals, and based on the total points earned by a project, the project could have its permitting fees waived, or waive up to ten years of incremental increase in property taxes. Although development that is not consistent with the city's smart growth strategy can still, and often does, occur, the expectation is that the financial incentives will act as a catalyst for developers to trend toward projects that benefit the city's environment. This approach is market-based rather than regulatory in the sense that developers can factor the added financial benefits from the property tax savings accrued by pursuing environmentally responsible development projects into their estimates of the returns they will make on their investments. To date, no comprehensive assessment of this voluntary approach has been conducted to try to determine whether, or to what extent, it has been able to produce development patterns with fewer environmental impacts, or even whether this approach is in fact more economically efficient.

Zoning and Comprehensive Land Use Planning

Perhaps the most ambitious of all efforts to engage in smart growth are those that are conducted in the context of the overall zoning and comprehensive land use plans of the city. Although many cities in the United States simply do not have much legal authority to determine the character of land use within their borders, many others do. Indeed, in some states, every city and town may now be required to engage in some sort of explicit land use or comprehensive planning process. As noted earlier, land use regulation, zoning, and growth controls have been the subject of numerous state and local ballot initiatives over the last few years. Many cities make land use the target of comprehensive strategic planning processes to try to influence the character of economic development looking up to five or even ten years into the future. Usually this takes the form of defining specified areas where economic growth and development can take place, including delineating urban growth boundaries. If the degree to which a city can be said to be sustainable is at least partly dependent on the nature of land use, particularly the density of housing and absence of sprawl, the delineation of open spaces, the degree to which there are specific mixes of land uses, and close proximity to public transit, and avoidance of environmental degradation, then zoning represents the local regulatory mechanism.

It is not clear how effective cities' land use and zoning strategies to control growth and development can be. Many cities around the country use their zoning authority to limit growth, and, as discussed later, this may turn out to be a futile exercise. Other cities, including many cities engaged in sustainability initiatives, instead attempt to use land use controls and zoning to channel or direct economic development and growth into specific areas of the city, or as a mechanism for picking and choosing which economic development would be most consistent with the quality of life being sought.

Just as zoning is used to determine what kinds of development activities can take place where, some cities are increasingly using zoning practices to delineate areas for special environmental protection. Perhaps the best example of this comes from Portland, Oregon, where the city's zoning code includes "environmental zones." Figure 4.1 shows a diagram of an environmental zone and its subzones as applied in the city. The environmental zone mainly consists of the resource area, shaded gray, and an area surrounding the resource area, called a transition area. Different land use regulations and development standards apply to each subzone. As the city's zoning ordinance stipulates, an "Environmental Protection zone provides the highest level of protection to the most important resources and functional values. These resources and functional values are identified and assigned value in the inventory and economic, social, environmental, and energy (ESEE) analysis for each specific study area. Development will be approved in the environmental protection zone only in rare and unusual circumstances" (City of Portland 2000). In addition, the city's zoning ordinance provides for "greenway zones" that afford additional protections mainly for setbacks from waterways.

Despite the use of zoning for this purpose, it is clear that the actual effects of these land use controls can be somewhat limited. In a study of the effect of statewide growth management policies in Oregon cities, Nelson and Moore discovered that, except in Portland, a significant amount of new development occurred outside of the designated growth boundaries, and that much of the new growth was of lower density than originally expected (Nelson and Moore 1996). This apparently happened

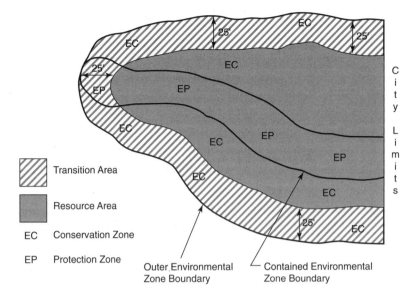

Figure 4.1
Diagram depicting an environmental zone, as defined in the City of Portland, Oregon
Source: City of Portland, Bureau of Planning. Chapter 33,430: Environmental Zones. Found at: ⟨http://www.planning.ci.portland.or.us/zoning/Zctest/400/430_Envir.pdf⟩.

for several reasons, including the designation of areas that were exempt from regulation. The fact that Portland was apparently more successful than other Oregon cities in managing growth may well be testament to how seriously the city seems to take sustainability. Yet even here, there is some question as to what the motivation of growth management is. In particular, there is concern that growth management efforts associated with Portland's sustainability initiative may have been motivated by the city growth machine's pursuit of economic development (Leo 1998). It should also be noted that, from a pure economic point of view, there may well be costs associated with such growth management. Another study of Oregon suggests that such management comes at the expense of the overall efficiency of the state's economy (Charles 1998). Nevertheless, growth management is rarely pursued for the purpose of promoting greater economic efficiency per se. It is pursued because it is thought to be one way of permitting economic growth in ways that minimize undesirable longer-term social and environmental impacts.

Another example of the aggressive use of land use planning to promote sustainable development also comes from the Pacific Northwest, the city of Seattle. Seattle's Comprehensive Plan Land Use Element is oriented around creating "urban villages" as its preferred development pattern. The idea behind this portion of the Comprehensive Plan is to establish land use regulations that encourage some kinds of development, discourage others, and prevent still others. As the 2000 Plan states it:

> The preferred development character is to be achieved by directing future growth to mixed use neighborhoods—designated as "urban villages"—where conditions can best support increased density. . . . Urban villages are intended to be a community resource enabling the City to deliver services more equitably, to pursue a development pattern that is more environmentally and economically sound, and to provide a better means of coping with growth and change through collaboration with the community . . ." (City of Seattle 2002, 1)

The preferred development plan is implemented through the zoning process, where permissible land uses are clearly outlined for the entire city. Despite the good intentions behind this approach, particularly the definition of urban villages, residents in specific neighborhoods in Seattle have not met it with universal enthusiasm (Hogen-Esch 2001). These are two examples of what cities that seem to take sustainability seriously have done as a matter of policy to regulate land use consistent with their sustainability visions. Yet even in here, it is not at all clear how effective they have been.

As part of the Seattle comprehensive sustainability land use plan, the city embarked on an effort to utilize Washington State's Growth Management Act in the creation of urban villages to accommodate new population growth. In 1992, Mayor Norm Rice unveiled a proposal to develop some 39 new urban villages by 2014, enabling the city to accommodate about 72,000 new residents (Iglitzin 1995). Urban villages, or ecovillages, represent attempts to design relatively high-density residential enclaves or areas that minimize environmental impacts by, for example, minimizing the need for automobile transportation. There is some evidence that, while there was initially a broad-based coalition in favor of the urban village approach, when the plans to develop specific areas of the city emerged, residents in those specific areas became vocal opponents. In the West Seattle part of the city, where the plan called for the development of four urban villages, existing residents' support apparently turned to opposition when they learned that these villages necessarily meant greater housing density. Moreover, they felt that the city was, in effect, seeking to diminish what they perceived to be their quality of life while leaving untouched the density of housing and development in more affluent parts of the city (Hogen-Esch 2001).

Eco-Industrial Parks

As discussed briefly in chapter 3, many cities have begun to look seriously at the prospect of engaging in economic development through the creation of eco-industrial parks. Cities such as Oakland and San Francisco, California; Plattsburgh, New York; Brownsville, Texas; Baltimore, Maryland; Atlanta, Georgia; Tucson, Arizona; and others, have begun to pursue such economic development in some way (Cornell University 2001; Smart Growth Network 2000a). Chattanooga is often cited as a leader in eco-industrial park development, but more than anything this presents a case study of how difficult it is to accomplish in practice what is intended in theory. The concept of an eco-industrial park, at least in practice, is ambiguous in the sense that it can mean at least two different things. On one hand, some cities apply the term to planned centers of industry where each separate industry located in the park is engaged in some sort of environmentally friendly activity. On the other hand, the "intellectual underpinnings" of eco-industrial parks suggest that there must be more than just environmentally responsible manufacturing processes or business practices. According to this second meaning, eco-industrial parks are centers of industrial activity that, unlike the more common industrial park, necessarily are linked together by their relationship to one another. Instead of having an industrial park with a dozen or more disparate businesses, and creating their resulting ecological footprints as a consequence of the material flows that are needed to support those businesses, an eco-industrial park tries to minimize those flows by placing cheek-by-jowl businesses that can use each other's products and by-products or wastes. Rather than having an industry that produces hazardous solvents as by-products of manufacturing ship those hazardous solvents out of town, why not locate a business nearby that can use those solvents in its own production processes? The idea is that the eco-industrial park creates a closed loop for the factors of production, maximizes industrial recycling, minimizes the flows of materials into and out of the city, and produces less opportunity for the city's residents to be exposed to possible environmental hazards, particularly in transit. Presumably, an eco-industrial park that conforms to the latter definition is more ecological friendly than one that conforms to the first.

Many cities in the United States are exploring the idea of developing eco-industrial parks, but few have made much progress in achieving specific results. Moreover, most cities that have considered eco-industrial parks seem to rely more on the first than the second definition, and this usually happens for pragmatic reasons. Chattanooga, for example, has at least two eco-industrial parks in the works, one that is referred to as the South Central Business District Eco-Industrial Park, and the other that is referred to as "The Volunteer Site." The intent of the former is to produce a "zero-emissions industrial facility," although very little has actually been accomplished there (City of Chattanooga 1995). The latter site consists of a former Army ammunition plant, about 7,000 acres in size, where much of the effort in redeveloping this site is focused on treating it as a contaminated brownfield. A primary goal of the effort is focused on finding ways of cleaning up the contaminated site, and of using the land for economically productive purposes in the face of liability for the cleanup. The site, which is operated by ICI Americas, has been designated by the city as a potential eco-industrial park. As explained by Sid Saunders, Vice President for Government Operations, "What that means to us is a community of businesses that collaborate to improve their economic and environmental performance by sharing resources like water, power, waste, heat, land, etc." As is true of many federal facilities, Saunders notes that "It is hard to get restoration/remediation money today, but leasing the site leaves past environmental liability with the Army and, at the same time, provides it with revenues needed to maintain the plant's standby status and to remediate the environmental problems from the past" (cited in Spohn 1997). To date, the Volunteer Site has not been successful in accomplishing much in the way of results, and indeed, most of the discussions and planning that have taken place around it center on pure economic development considerations (Chattanooga Chamber of Commerce, 2000; Smart Growth Network 2000b).

Private Sector Involvement (Public–Private Partnerships)

One of the omnipresent undertones surrounding sustainable cities initiatives concerns the role of the private sector. In some cities, sustainability initiatives are pursued completely independent of the private sector and business community. In places such as Seattle and San Francisco, the sustainability initiatives appear to have little or no direct involvement of the business community, per se. Presumably, cities that elect not to involve business and industry are suspicious of what including the private sector might do to sustainability, perhaps fearing that business will be less than fully supportive of the actions that need to be taken to move toward becoming more sustainable. Alternatively, they may discover that, for whatever reason, business and industry leaders simply do not wish to participate in such processes, particularly if the initiative has strong "proenvironmental" overtones. In other cities, an explicit effort is made to find ways of including business and local industry. These cities seem to have concluded that if progress toward sustainability is going to be achieved, then the local private sector needs to be involved. Places such as Boulder, Chattanooga, and to a lesser degree, Jacksonville, have carved out specific roles for the business community. Very often, this role takes the form of some type of public–private partnership.

Boulder, Colorado, has created such a partnership. The Boulder Partners for a Clean Environment, or PACE, program represents a voluntary, nonregulatory program that offers free pollution prevention education and technical assistance to Boulder County businesses. PACE is a cooperative effort of the city of Boulder Office of Environmental Affairs, Boulder County Health Department, city of Longmont, Boulder Energy Conservation Center, and Boulder Chamber of Commerce. A great deal of program support has been provided by the city of Boulder Water Quality and Environmental Services, and the City of Longmont Water/Wastewater Utilities beginning in 1998.

As discussed in chapter 3, PACE focuses on small- and medium-sized businesses that are typically too small to be regulated by the state, and may not have the time and resources to investigate and implement pollution prevention measures on their own. Sectors are prioritized and targeted based on the environmental impact of the business sector in Boulder County and the availability of pollution prevention options. In brief, PACE works with small businesses to create a set of industry-specific criteria and standards for certification as PACE-compliant. In 1997 and 1998, the program focused on auto repair, auto body, and print shops. In general, these shops are exempt from air and/or water permitting regulations, but collectively they use significant quantities of hazardous materials for which pollution prevention alternatives exist. These shops may also be unaware of their requirements for hazardous waste management, and PACE program staff provide a tremendous service in providing compliance information. PACE developed certification criteria for manufacturers, dental offices, and restaurants. Outreach to these sectors began in 1999. At last count, over 200 local printers, auto repair shops, dentists, and restaurants had either been certified, or were working toward certification. The program claims to have achieved impressive reductions in pollution, particularly in volatile organic chemical air emissions (PACE 2000).

Other forms of public–private partners exist as well. In Chattanooga, such partnerships frequently take the form of nonprofit economic development corporations. Considered by many in the city to be a central element in its sustainability initiative, the RiverValley Partners engaged in a wide array of economic development projects and activities. Originally funded by eight banks and seven foundations, this organization apparently created something of a symbiotic relationship between the city government and the private sector, including the Chamber of Commerce, to help contribute to sustainable economic development (Beatley and Manning 1997, 164). Projects such as the Tennessee Aquarium and the Chattanooga Riverwalk, ostensibly motivated by an interest in creating economic development within the context of the city's quality of life visioning process, were the direct products of this partnership. Recently, however, the name of this initiative was changed and it was swallowed up by the Chattanooga Area Chamber of Commerce, and is now a small part of that organization's regional growth initiative. Although it is too soon to tell whether the balancing associated with the RiverValley Partners has been overwhelmed by the pursuit of pure economic development, the language of the Chamber suggests that the central concern is for marshalling all available resources to be used strategically for that purpose (Chattanooga Chamber of Commerce 2001).

Green Building Programs

As noted in chapter 3, many cities have tried to make conscientious efforts to adapt existing building construction regulations to foster greater energy efficiency, and to reduce environmental impacts. This includes reducing impacts produced during and after construction, as well as impacts produced in the manufacture of specific building products. Such "green building" programs have been developed all around the country, including a recently adopted program in Scottsdale, Arizona (City of Scottsdale 2001). Most green building programs focus on how new buildings are designed and constructed, although they can also focus on how old buildings are "desconstructed" or torn down. Not all such programs affect all kinds of new buildings, with some focusing only on municipal buildings and construction, others on residential construction, some only on new homes, some only on commercial construction, and so on. Those that do include commercial construction probably have the greatest overall effect on local economic development.

In 1999, San Francisco enacted an ordinance that places fairly strict regulations on the design and construction of municipal buildings. This ordinance requires that specific attention be given to resource efficiency and conservation, indoor environmental quality, materials efficiency, occupant health and productivity, transportation efficiency, minimized use of toxic materials and minimized production of hazardous waste, deterrence to pest infestations, and reduced impact on ecosystems. Because this ordinance is limited to the development actually done by the city government itself, the effect on overall economic development is also limited. Some cities, however, have developed green building programs that are far more expansive.

In Santa Monica, the city has focused mainly on commercial development—institutional and commercial offices, light industrial buildings, commercial retail buildings, multifamily residences, and hotels and motels. It uses a performance-based standard, seeking a 20 to 25% energy conservation target, to require a number of features in all new construction of this type. The city's official information describes the program by saying that:

In September 1994 Santa Monica's City Council adopted the Santa Monica Sustainable City Program. This program was developed by the city's Task Force on the Environment as a way to create the basis for a more sustainable way of life—one that safeguards and enhances local resources, prevents harm to the natural environment and human health, and strengthens the community and local economy—for the sake of current and future generations. Conventional design and construction methods produce buildings that can negatively impact the environment as well as occupant health and productivity. These buildings are expensive to operate and contribute to excessive resource consumption, waste generation, and pollution. To help reduce these impacts and meet the goals of the Sustainable City Program, the Task Force recommended that the City adopt a set of guidelines to facilitate the development of "green" buildings in Santa Monica without forcing excessive costs or other burdens upon developers, building owners or occupants. The Green Building Design and Construction Guidelines were developed over a three-year period with extensive input from the local design, construction and development community. The Guidelines include required and recommended practices that are intended to reduce life-cycle environmental impacts associated with the construction and operation of both commercial and municipal developments and major remodel projects in Santa Monica. They provide specific "green" design and construction strategies in the following topic areas: Building Site and Form, Landscaping, Transportation, Building Envelope and Space Planning, Building Materials, Water Systems, Electrical Systems, HVAC Systems, Control Systems, Construction Management, and Commissioning. They are not intended to address development of single-family residential dwellings and duplexes, high rise buildings, or occupancies with special process demands (heavy industrial operations, car washes, service garages, etc.), however many of the recommended practices presented in the Guidelines are relevant to these building types as well. (City of Santa Monica 2002)

The city's two ordinances require that new construction utilize building practices that are designed to improve performance in seven environmental areas: water runoff and irrigation; bicycle storage; storage space for, and use of, recycled materials; water conservation; energy conservation; wastewater restrictions; and management of new construction itself. Beyond these areas where developers are required to meet performance standards, there are also numerous recommended design features and approaches contained in the green building program.

Boulder uses a different approach. Boulder's "Green Points" building program provides technical information to designers, architects, and homeowners, but the heart of the program is found in its tie to the permitting process. In order to obtain a building permit, the design of new construction over 500 square feet must demonstrate compliance with environmentally sound building practices. It does this by earning points, which are assigned to the design features of the new construction. Points are assigned for environmentally friendly land use, framing, plumbing, electrical, insulation, heating, ventilation, and air conditioning, use of solar, and indoor air quality. Use of specific building techniques and/or materials are assigned a specified number of points, and the overall design plan must earn a total of between 10 and 25 points depending on the size of the new construction. For example, a new building of 2,500 square feet would have to earn 25 points, and it could do so by using engineered lumber in the roof or floor (5 points), using a tankless hot water system (2 points), using light fixtures with compact fluorescent bulbs installed (1–3 points), installing R-24 wall insulation (3 points), using radiant floor heating (5 points), and so on. Presumably, the result of these regulations is to create an ever-growing housing stock that is increasingly energy efficient.

Support for Local Commerce

As noted earlier, Kline (1995) suggests that one of the tenets of smart growth is the notion that efforts need to be made to keep economic resources local. Just as efforts to minimize the ecological footprint of the city require "closing the loop" for wastes and other environmentally impacting products and processes, smart growth tries to close the economic loop by keeping locally generated capital within the city's borders. This applies to neighborhood reinvestment, and it also applies to efforts to provide increased opportunities to support local commerce. One of the elements of Olympia, Washington's sustainability effort is found in the Sustainable Community Roundtable's local currency program called "Sound Exchange" (City of Olympia 2001). The idea behind this effort is to create a local currency program that provides new opportunities for consumers to keep their money circulating within the city, or to at least affect the local "balance of trade." The way it works is that local businesses agree to honor "Sound Hours," the unit of currency, which is ostensibly worth one hour of labor, or about $10. Consumers can purchase these Sound Hours, and use them as currency only at businesses within Olympia. Although it would appear there are few economic incentives for consumers to purchase Sound Hours (there does not appear to be any sort of discount, for example, when purchasing a Sound Hour, and the price of using a Sound Hour may actually be slightly higher), at last count the program involved some 500 local businesses and had achieved a circulation of about $25,000 a year.

Critiques of Smart Growth Approaches

While smart growth approaches sometimes try to channel growth into particular areas or to imposed greater selectivity on the type of economic development that takes place, other smart growth efforts simply try to slow down the rate of growth. The idea that a city, local government, or even the local governments in a region can effectively limit growth has been the subject of some debate. Indeed, a prominent view is that such measures will inevitably fail to achieve their goals, and may make problems of sustainability even worse. Commenting on some of the efforts by cities to control growth or engage in smart growth, with particular reference to Denver and Phoenix, Downs expresses his skepticism:

> In my opinion, individual local governments cannot influence the overall growth rates of their regions or states by implementing local growth-control laws, statewide initiatives notwithstanding. Regional growth rates are determined by broad forces beyond the purview of any one or even several local jurisdictions. . . . So when one locality passes laws limiting future growth within its own boundaries, it does not affect the future growth of the overall region, but rather moves the region's future growth to other localities, or to outlying unincorporated areas. Local growth limits, then actually aggravate sprawl." (Downs 2000, 4)

Downs offers the advice that local governments ought to work to create incentives for development to take place in particular areas of their cities—areas that are thought to create fewer negative impacts on the city—rather than trying to limit growth altogether. Although the cities' sustainability efforts encompass a wide array of economic development approaches, many of them do in fact seek to direct economic development in that way. Smart growth and growth limits have also been criticized as being elitist and, as discussed in chapter 6, as being inequitable on racial, ethnic, and economic grounds.

Summary: Sustainable Approaches to Economic Development

Unlike advocates of no-growth, sustainable growth proponents seek to find ways of managing the form and location of economic development in their cities. The expectation is that by doing so, there will be two related consequences. First, the city will be able to exercise some control over the biophysical environment by moderating the environmental degradation or damage that would be produced by unfettered economic development. Cities are often able to keep out business activities that pollute, or that extend the city's ecological footprint. Cities sometimes are able to intervene to assist businesses to find ways of shrinking their ecological footprints by helping them pollute less. Cities are able to effectuate programs, such as green building, that reduce local energy consumption. As with eco-industrial parks, cities are able to encourage related industrial activities that in the aggregate produce a smaller ecological footprint than would otherwise be produced. Second, the city will be able to try to create or maintain a particular level and type of livability—the quality of life that results will be more in tune with what the residents of the city want. Smart growth cities clearly believe that the consequence of efforts to moderate environmental damage is that they become more livable places. Such cities are able to manage a whole array of quality of life factors, including traffic congestion, the quality of housing, quality of jobs and employment, the amount of capital that stays local, and others.

Clearly, cities that take sustainability seriously bring a very different mind-set to economic development than cities that do not. Not all cities are as free to pursue smart growth kinds of approaches to economic development as they might like, but the desire to encourage some kinds of economic development rather than others is clearly present. In cities that are most aggressive, the full regulatory and enforcement power in brought to bear, as in green building programs that establish requirements for obtaining building permits. Aggressive efforts to tackle brownfield redevelopment constitute another smart growth activity characteristic of cities that take sustainability seriously. It may be demonstrable that smart growth approaches to sustainable cities produce less economic growth than unregulated approaches to economic growth, but that is not the issue for sustainable cities. Sustainable cities certainly believe they can achieve an adequate and sustainable level of economic growth and development while at the same time safeguarding their environments and their quality of life. Whether all these smart growth economic development activities actually produce or help produce a sustainable city remains to be seen.

5

Communitarian Foundations of Sustainable Cities: A Solution to the Tragedy of the Commons, the NIMBY Syndrome, and Transboundary Impacts?

What is evident from many pragmatic definitions of sustainable communities, particularly those that emerge from cities' sustainable indicators projects, is that there is often deep concern for elements of human interaction and relationships. In many cities, the concept of sustainable communities is not just about protecting the environment, or controlling economic growth for the benefit of the environment. It is also about the relationship between the physical environment and the people who populate it, including a wide range of social issues that transcend the purely environmental. This view of sustainable cities places great import on the function of civil society—the institutions and social processes in society that influence how residents interact (or do not interact) with each other. Thus, in many respects, it is also about how the environment is defined in the sense that many believe it is desirable to move beyond a conception of the environment in purely biophysical terms to a view that is more expansive and inclusive. But this raises a variety of issues and questions. Given the underlying conceptual underpinnings for sustainability, per se, is there reason to pursue an expanded definition of what constitutes the environment? Is there some expected relationship or causal connection between the biophysical environment and the function of civil society? Is the character of the local civil society in some way a determinant of or influence on whether cities will choose to be aggressive in pursuing sustainability? Can cities' sustainability initiatives, in practice, accommodate expanded definitions that incorporate issues of civil society?

The purpose of this chapter is to examine what might be called the pragmatic and conceptual "communitarian" underpinnings of the concept of sustainable cities, ultimately arguing that the concept of sustainable communities, as originally imagined and articulated by some, and as practiced in many cities, may well be about finding ways to pursue economic growth while explicitly promoting particular kinds of human interactions and shaping civil society. This is based on the notion that explicit attention needs to be paid to the relationship between economic growth and protecting the environment on one hand and what it takes to make cities livable on the other. To many advocates of sustainable communities, making cities livable requires changing the fabric of civil society. To state it succinctly, the concept of sustainable communities is, for many cities and in many respects, fundamentally communitarian in nature. Without changing the way people relate to each other, and the values that underlie these interactions, pursuing sustainability would simply not be possible. Although this is an idea that is generally still in search of empirical support, there is little question that in a number of cities, there are clear assumptions about the nature of this relationship. It is also clear that many advocates of sustainability see an explicit connection between the condition of the environment and the operation of civil society, whether or not cities' particular sustainability initiatives endorse it.

At the same time, however, it is the communitarian character of sustainable cities that creates an enormously difficult challenge. Just as many conceptions of sustainability rely on communitarian principles, it is these very principles that make the pursuit of sustainable cities a very political process. In many cities, particularly where the dominant culture is something other than communitarian, sustainability initiatives encounter obstacles from political and civic leaders. Very often, this clash represents the tension between pursuing traditional economic development and following a different, more environmentally friendly, path. As is developed later, many cities' sustainability projects are initiated by people who believe that such projects will unleash a populist, neighborhood-based, assault on the dominant political, economic, and social values. Indeed, many local advocates of sustainability see the sustainable indicators project potentially as the vehicle for producing a wide array of social change. When the dominant social and political values turn out to be far more resistant to change than anticipated, when such projects experience little success in getting communitarian issues on the public agenda, interest in them sometimes wanes.

In chapter 2, as part of the effort to examine ways of measuring how seriously cities take the pursuit of sustainability, it became clear that there is often a prescription calling for greater public participation and participatory activities, and that the pursuit of greater public involvement constitutes an important and integral piece of how some cities define sustainability. Cities' sustainability initiatives themselves rarely spell out any sort of reason why public participation, "empowerment," and community building and other dimensions of civil society are thought to be important elements of their efforts. Even so, there is a clear logic that prescribes why concerns for community might be said to be integral components of any initiative that takes sustainability seriously. Indeed, one of the features that helps distinguish cities' sustainability efforts centers on the extent to which such efforts actually seek to promote citizen participation and involvement, interpersonal interactions directed at improving the operations of government, and community-based problem solving.

Whether or not one considers these elements of sustainable communities to be appropriate depends on how one defines a sustainable community in the first place. Given the conceptual literature on sustainability, it is difficult to avoid the conclusion that sustainable cities can contain a core of fundamentally communitarian concerns. Not all advocates of sustainable communities accept the idea that there must be, or ought to be, a heavy communitarian element. Indeed, as discussed later, even among many advocates of sustainable cities, there is a strong suggestion that pursuit of communitarian goals gets in the way of pursuing environmental goals. This represents yet another side to the challenge of pursuing sustainability in cities. The decidedly communitarian elements of sustainable cities definitions are often placed at loggerheads with more strictly environmental definitions. Many advocates of sustainable cities argue that the pursuit of sustainability is first and foremost about protecting and improving the physical environment. To such advocates, many of the social and communitarian elements of sustainability are subordinate or irrelevant to the principal task at hand.

Many proponents of sustainability see the pursuit of sustainability as a highly technical and "professional" endeavor where there is no need for large-scale public involvement. To these advocates, there is no question what needs to be accomplished, and there really is no difference of opinion about what environmental problems need to be addressed and how they should be addressed. If a city has an internal air pollution problem, so the argument goes, correcting this problem is a job for professionals (Brugmann 1997). One does not need to build a community or to create a communitarian utopia to redress the essential problem. Thus, in the view of many advocates of sustainable cities, priorities are misplaced.

The essential problem confronted by communitarian conceptions of sustainable cities is the idea that sustainability can be pursued as a matter of elite and professional preferences. If air pollution is a purely technical problem, then why have we not corrected the problem years ago? If water pollution is a matter for the professionals, then why are our waterways and groundwater supplies still threatened? For many observers, the reason is fundamentally one of public values—the value that political and business leaders and the general public place on individual freedoms and liberties. As long as most people are willing to accept the status quo, little progress toward sustainability is possible. In this argument, the *political will* to pursue sustainability prevents all those professionals and technical experts from doing their part. So the question becomes, how can this political will be stimulated? To communitarians, until the political will to pursue sustainable cities has been developed, marginal improvements in sustainability and livability are the best that can be achieved. Skeptics such as Ophuls and Boyan (1992, 237–253) doubt that even marginal improvements are possible. Communitarians offer the hope that addressing issues of civil society can begin the process of altering community values, and make the pursuit of sustainability possible.

To the extent that communitarian values are found in sustainable communities conceptions, this means that pursuing sustainable communities can be just as much about building communities of people as it is about achieving sustainable economic development results or protecting the environment. This chapter examines this issue in more depth, delving into the conceptual reasons why this is thought by some to be such an important part of achieving sustainability. As such, clarification of these communitarian components of sustainable cities will set the stage in this and later chapters for the empirical exploration into prerequisites and consequences of efforts to achieve sustainable communities.

There are plenty of conceptual and practical reasons to think that sustainable cities are unattainable. Many of these reasons have been discussed earlier. Suffice it to say here that the city, or any other relatively small geographic area, may not necessarily represent the ideal level at which to pursue sustainability. However, there are reasons to believe that cities and communities offer greater opportunities for taking sustainability seriously, and many of these reasons have to do with the role and necessity of particular kinds of interpersonal interactions, especially interactions that are related to behavior toward the environment, consumer behavior, and perhaps most important, democratic governance behavior along with the attitudes and values that underlie them. If social values need to be changed in order to embrace the basic tenets of sustainability, then the way communities of people operate stands to play a significant role in making progress toward sustainability.

The Communitarian Foundations of Sustainable Communities

Why are the communitarian elements considered by some to be important, if not essential, components of sustainable cities? Why do some people believe that community building and participatory processes are, or should be, such an integral part of taking sustainability seriously? The starting point for understanding the communitarian foundations of sustainable communities must be recognition of the "three deadly sins" often thought to impede progress toward sustainability, three generalizable social and governance problems that are thought to lie at the heart of *un*sustainability. These three problems are the "tragedy of the commons," the "Not-in-my-backyard (NIMBY) syndrome," and the expansion of cities' ecological footprints that result from "transboundary shifting" of environmental impacts. Each of these phenomena contributes to the difficulty of making small geographic areas sustainable in the sense that they represent system-level consequences of individual-level attitudes, values, and behaviors. And it is these three deadly sins that communitarian approaches to sustainability promise to redress.

All three of these problems are fed by what some commentators call "rampant individualism," where individuals are free to act on what they believe to be their own immediate self-interest. The basic argument concerning these problems is that they are created as a result of a mismatch between what is good for society, or in this case, the community, and what individual people think is good for them personally. Contrary to the basic tenets of neoclassical economics and political liberalism, the communitarian view suggests that what is good for the community in the aggregate is not always simply the sum total of what is good for each of the individuals acting as "rational consumers" in that community. In short, the communitarian view of sustainability suggests that there really is no functioning mechanism in communities today to encourage individuals to consider the community or aggregate consequences of their imputations of self-interest and the personal behavior that stems from them.

Although it cannot be denied that individualism in some form constitutes a core value on which the United States was founded, recently thinkers have emphasized the idea that alternative foundations were also present. One of these alternative foundations, very much a part of the history and culture of the United States, focuses on communitarian rather than individualist values. In other words, U.S. history does not provide an unfettered tradition of libertarian values, but rather it provides an understanding of individual freedoms as being constrained by community concerns. This is an issue that itself is far too broad to address here in detail. However, what makes this communitarian foundation so important to understand is that, if rampant individualism constitutes a value that feeds the three problems associated with unsustainability, then communitarianism promises to offer a conceptual solution to these problems.

The need for sustainable cities initiatives, from a communitarian perspective, has never been greater. The values and attitudes that support and reinforce unsustainable practices have become stronger, and the mechanisms that support and reinforce shared values and understandings of commons issues have become weaker. Witness, for example, the growth of "green backlash" initiatives, such as the Sagebrush Rebellion of the 1980s, and the more recent New Property Rights movement. These represent strong political initiatives aimed at undermining communitarian solutions to environmental protection (Switzer 1997). Underlying the link between the need to build community and the pursuit of sustainability is the notion that, for a variety of reasons, the mechanisms that could once have been counted on are disappearing. The local political organizations and organizations of civil society that once kept residents personally engaged, active, and edified have diminished and declined, replaced by large national lobbying organizations that do not typically perform these functions. When Robert Putnam (1995, 2000) argues that we are now "bowling alone," and when the National Commission on Civic Renewal (1998) calls us "a nation of spectators," they are making the argument that the loss of opportunities for people to interact with each other has undermined the creation of shared values and understandings. Applied to sustainable cities, this means that the key problems, the tragedy of the commons, the NIMBY Syndrome, and resulting transboundary problems, are getting worse. As the institutions and organizations of civil society have declined, so the argument goes, there is no longer any social or political mechanism to mitigate the rampant individualism that contributes to unsustainability. As we will see, cities that take sustainability seriously tend to seek to redress this issue through the creation of initiatives that are designed to engage residents and build a shared understanding and vision of the city and its environment.

Sustainable Communities as a Communitarian Concept

Just as the idea of sustainable communities grew in part out of a particular understanding of sustainability, so too did it grow out of a particular understanding of "community," or more specifically the role of "community" in contributing to a more sustainable environment. Often, critics of the new localism of environmental policy misunderstand the conceptual importance of community to sustainability, or dismiss the importance of community as having no empirical support, or worse, distrust community as the last bastions of parochialism. Communitarian thought nonetheless provides perhaps the strongest conceptual underpinnings to the idea of sustainable communities. There is no doubt that underlying much of the advocacy of sustainable cities and sustainable communities is the notion that associated participatory processes will become instrumental in transforming cities into environmentally responsible places. Stated in Putnam's terms, the processes associated with creating and maintaining a sustainable city are expected to fill the social capital void left by decades of benign neglect to issues of interpersonal interactions. Advocates of sustainable communities seem to think that promoting such interactions will create the political foundation and support necessary for the pursuit of environmentally responsible local policies.

Much of the contemporary communitarian movement is largely founded on the notion that liberalism has fostered the evolution of communities without shared values. Liberalism's focus on individualism, according to this view, has produced a shift "in our practices and institutions, from a public philosophy of common purposes to one of fair procedures, from a politics of good to a politics of right, from the national republic to the procedural republic." (Sandel 1984, 183) As Barber states it, liberalism has created a political system that "can conceive of no form of citizenship other than the self-interested bargain"(Barber 1984, 24). The result is that citizens increasingly see the personal gains, particularly short-run gains, they can accrue from acting as individuals, but do not see how the impacts they produce affect others.

Perhaps equally important, the institutions of society—the components of civil society—that could conceivably contribute to the development of shared values and promote broader understandings of individuals' impacts on others, have been all but forgotten in contemporary America. The communitarian movement seeks to reverse this trend by reasserting the importance of shared community values. Thus, when Etzioni (1996) writes that "the virtue of *stewardship toward the earth*, the commitment to the environment as a common good, is profoundly communitarian, on the face of it," (252) he is asserting a connection between communitarian values and sustainability. This connection may seem obvious to some, yet it is precisely this communitarian element that is often missing in many applied sustainable communities efforts, and often denigrated as being irrelevant or counter to local sustainable *economic* development. As is discussed later, many cities that purport to take sustainability seriously do not seem to embrace the idea that the environment is a common good except to the extent that it contributes to local economic development in pretty mainstream ways. Nevertheless, it is from these basic communitarian foundations that prescriptions for emphasis on community building and participatory processes emerge (Press 1994).

Civic Environmentalism and Community

Much of the foundation for the communitarian content of sustainable communities may be found embedded in the concept of "civic environmentalism." This concept, which has developed since the early 1990s, represents an effort to offer alternatives to, or more accurately, complements to, federal and state command-and-control environmental regulation. The idea, as articulated by DeWitt John (1994), is that there is much that local communities can do to improve and protect the environment even when federal and state regulatory agencies are legally or politically constrained from acting. To John, civic environmentalism grew out of the frustrations with environmentalists during the 1980s when environmental protection was clearly a low priority in federal policy. As John notes, ". . . in some cases, communities and states will organize on their own to protect the environment even without being forced to do so by the federal government. . . . Civic environmentalism is fundamentally a bottom-up approach to environmental protection." (John 1994, 7)

But civic environmentalism has evolved beyond simply prescribing a role for communities in environmental protection to incorporate many communitarian notions of participatory processes. John offered an initial sense of this when discussing the processes that he associates with locally based environmental protection when he notes that:

Civic environmentalism . . . tends to involve a different style of politics than command-and-control regulation. There are still strong differences of opinion [in environmental protection decisions], but there are fewer confrontations between black hat polluters and white hat protectors of the public trust, and there is more bargaining among a diverse set of participants. Civic environmentalism is a more collaborative, integrative approach to environmental policy than traditional regulation. (John 1994, 10)

In John's conception, it is not clear how central is the role of participatory and collaborative processes. Clearly the kinds of decision-making processes he has in mind are participatory, but it is not clear whether they necessarily must be. As the concept of civic environmentalism has evolved, however, participatory processes have become much more central. Much like the notions put forth by communitarians, civic environmentalists suggest that participatory processes are necessary to build the social capital required for the effective pursuit of sustainability (Potapchuk 1996). As Selman and Parker note with reference to the goals of the Local Agenda 21 process in the United Kingdom, there is some expectation that widespread participation in local environmental planning and programs will build "better" citizens, better consumers, and a more environmentally conscientious populace (Selman and Parker 1997). This is an idea that is also present in numerous other works (Agyeman and Evans 1995; John and Mlay 1999). However, virtually all of the advocates of creating more environmental communitarian places embrace the expectation that the quality of decisions, both personal and institutional, will be different as a result. This can be seen with specific reference to some of the major impediments to local environmental protection. Starting with the problem of the tragedy of the commons, participatory processes and civic environmentalism are thought to offer solutions.

The Problem of the Tragedy of the Commons
A central concern of communitarians' views of individualism is what is often referred to as the "tragedy of the commons." The tragedy, as described by Garrett Hardin (1968), occurs when commons goods or common-pool resources (such as the air and sometimes the water and public lands) are used by individuals for personal gain. The most frequently cited example of the tragedy is in the context of a group of cattle owners who share a parcel of grazing land. Each cattle owner is free to purchase as many cattle as can be afforded, and fatten up the cattle on the shared grazing land. Unfortunately, when all of the cattle owners pursue the same logic and strategy, the common resource, the grazing land, becomes depleted so that the cattle cannot thrive, and eventually cannot be sustained. No one cattle owner can decide to cut back on the number of cattle without making a sacrifice relative to the other cattle owners. In other words, if one cattle owner cuts back on the number of cattle, the other ranchers, acting rationally, would then be free to add more cattle. Not only does the owner who cut back lose out, but the other owners reap the windfall. But the consequence of the rational individual choice, the tragedy, is that all the cattle owners eventually lose

their cattle because the cattle collectively deplete the supply of food faster than it can be grown. This paradox is one that applies to a variety of types of common-pool resources. The problem is that there is usually no incentive for individuals, acting purely in pursuit of the short-term, self-interested bargain to use less air or water. To the contrary, in the absence of aggressive regulation, the incentives usually motivate the depletion of such commons goods.

Of course, neoclassical economics prescribes that markets themselves will correct this problem. Markets serve as self-correcting mechanisms where supply and demand adjust according to the depletion of natural resources. If the problem underlying the tragedy of the commons is that there is economic incentive for the cattle ranchers to deplete the grazing resource, so the argument goes, the market will adjust to remove this incentive. As the supply of cattle decreases because of the inability of the land to support them, the price of the cattle will increase to a point where prospective purchasers of the cattle will seek alternatives. This, in turn, lowers demand for the cattle, which causes the price of the cattle to fall. When the price of the cattle falls, the economic incentive to add more cattle to the grazing land, and the incentive to deplete the natural resource, is removed. Of course, the problem in the case of natural resource depletion is that often the environmental damage done by the time markets respond can be irreparable. While the vegetation in a grazing area may take a relatively short time to recover, analogous depleted resources may not recover as quickly or as completely. How long would it take the depleted fish stocks of the North Atlantic to recover in the absence of government intervention? How long would it take for market adjustments to undo the climate change effects of air pollution, and how much other damage would occur while these market adjustments take place? In short, it may not be possible or desirable to wait while market adjustments take place, if they take place at all.

In their scathing attack on dominant political and economic values in *Ecology and the Politics of Scarcity Revisited*, Ophuls and Boyan argue that under historical conditions of abundance of natural resources, individualism and the neoclassical economic order based on individualism make sense. Under conditions of increasing scarcity of natural resources, alternatives must be found, particularly in making commons decisions that are imperfectly governed by markets. In their advocacy of creating what they call a "steady-state society," a concept akin to sustainability, Ophuls and Boyan (1992) put forth the prescription that "rampant individualism" must be displaced by greater communalism. As they state it:

> The self-interest that individualistic political, economic, and social philosophies have justified as being in the overall best interests of the community . . . will begin to seem more and more reprehensible and illegitimate as pollution and other aspects of [natural resource] scarcity grow. And the traditional primacy of the community over the individual that has characterized virtually every other period of history will be restored. How far the subordination of individual to community values and interests will have to go and how it will be achieved are for the future to determine. (p. 285)

Although Ophuls and Boyan provide little in the way of explanation for *why* "community" is the appropriate level at which to begin achieving the steady state, the implication is clear that the political and economic order must be more attuned to needs of the *community*. Their definition of community, like that of many communitarians, stresses the importance of identifying *shared* community values, and ensuring that peoples' behavior is consistent with those values.

Numerous writers have addressed the issue of how to overcome the tragedy of the commons as it affects the environment, or at least have tried to understand the conditions under which people will actually make collective decisions that benefit the entire group. The picture they paint is not quite so bleak. Elinor Ostrom has examined such issues "in the field," studying actual small-scale, common-pool resources decisions around the world. In commenting on the rational choice expectation that common-pool resource decisions will virtually always end up in the tragedy of the commons, she suggests that "When individuals who have high discount rates and little mutual trust act independently, without the capacity to communicate, to enter into binding agreements, and to arrange for monitoring and enforcing mechanisms, they are not likely to choose jointly beneficial strategies . . ." (Ostrom 1990, 183). But she successfully identifies numerous cases where people do in fact make collectively beneficial decisions concerning the allocation of common pool resources. The stipulated conditions conducive to creating the tragedy may apply only to large-scale resource problems, such as decisions made by nations. On smaller scales, these conditions may well be violated or varied, and the result can often be mutually beneficial decisions. In case studies, Ostrom and colleagues conclude that "Simply allowing individuals to talk with one another is sufficient change in the decision environment to make a substantial difference in behavior . . . individuals . . . not only come to agreements but craft their own rules and enforce these rules without relying extensively on external authorities" (Ostrom, Gardner, and Walker 1994, 320). This idea certainly begins to add credence to the possibility that, at least in small scale common-pool resource decisions, establishing "community" could be critical, since establishing community is usually defined in terms of interpersonal interactions, including talking with one another.

The idea of a community basis for achieving some form of sustainability has been developed by a number of theorists. In his investigation into the social and political mechanisms that would promote environmental protection, John Dryzek's *Rational Ecology* (1987) proposes a "radical decentralization" of social choice to the local or community level. His explication clearly outlines the promise of community-based social choice when he states,

Local self-reliance . . . means, first and foremost, that communities and their members must pay great attention to the life-support capacities of the ecosystem(s) upon which they rely. This attention is not a matter of choice, other than the choice between life and death. Self-reliance bars access to—and exploitation of—distant "invisible acres," and rules out despoliation followed by emigration. Local residents are forced to heed negative feedback signals from their natural environment. Such signals have an immediacy and clarity which is generally lacking for most members of contemporary industrial societies (however "open"), who have scant knowledge of—let alone concern for—the ecosystems upon which they depend. (p. 218)

Thus, Dryzek sees something almost automatic about community-based social choice producing results that are less likely to foster commons tragedies. Clearly, communities seem to constitute a level at which political and economic decisions could be most friendly to the environment. At least that is the rationale put forth by Dryzek. Of course, the common wisdom is that this promise becomes dashed against the shoals of local politics and local economic development. This issue is addressed more thoroughly in the case examples of cities.

The economist Herman Daly and the theologian John Cobb (1994) provide perhaps the most systematic alternative to neoclassical economics that addresses the linkage between issues of sustainability and conceptions of community. In *For the Common Good*, a book best known for its sustainability prescriptions about neoclassical economics and advocacy of altering our system of national income accounting to internalize environmental quality and degradation, Daly and Cobb tackle the issue of individualism and self-interested economic behavior. Arguing that neoclassical economics' reliance on individualism leads to social results that are profoundly unsustainable, and recognizing the fact that their revised economics needs to transcend neoclassical definitions of self-interest, they develop an economics that accounts for the social and political community context in which individuals live and make their decisions. Just as neoclassical economics uses the individual person as the building block for a system of free enterprise, Daly and Cobb describe what they call the "person-in-community" as the fundamental building block of sustainability. In contrasting classical views of individualism with their revised view, Daly and Cobb (1994) write:

But what is equally important for the new model [of person-in-community]—and absent in the traditional [neo-classical] one—is the recognition that the well-being of a community as a whole is constitutive of each person's welfare. This is because each human being is constituted by relationships to others, and this pattern of relationships is at least as important as the possession of commodities. These relationships cannot be exchanged in a market. They can, nevertheless, be affected by the market, and when the market grows out of control of a community, the effects are almost always destructive. Hence, this model of person-in-community calls not only for provision of goods and services to individuals, but also *for an economic order that supports the pattern of personal relationships that make up the community.* [emphasis mine] (pp. 164–165)

Thus, the Daly and Cobb view of sustainability not only prescribes that it is something that should take place in the context of communities, but also that it must operate in support of the patterns of human interactions and relationships that exist in those communities. The particular kinds of relationships that Daly and Cobb prescribe as important for sustainability look remarkably communitarian in nature. After reviewing numerous definitions of "community," and discussing the elements that they consider essential to sustainability, Daly and Cobb suggest that there must be a significant communal character. In their words:

To have a communal character . . . does not entail intimacy among all participants. It does entail that membership in the society contributes to self-identification. We accept this requirement and add three others. A society should not be called a community unless (1) there is extensive participation by its members in the decisions by which its life is governed, (2) the society as a whole takes responsibility for the members, (3) this responsibility includes respect for the diverse individuality of these members. . . . The extent to which a society contributes to the identity of its members, the extent to which they participate in its governance, the extent to which it takes responsibility for its members and the extent to which it affirms them in their self-determined diversity all vary. . . . [but] we favor societies at many levels becoming more communal by all four of the criteria . . . (p. 172)

Clearly, the argument is that the function of the local economy must be put in service to the tasks of promoting a wide range of civic and political participation goals, and promoting the assumption of community responsibility and respect for diversity. These goals are decidedly communitarian in nature, concerned as much about the processes that are used to make public decisions as about the end results. The Daly and Cobb call for "extensive participation" and for the members of society to "take responsibility" for each other is central to communitarian views of governance. The tragedy of the commons is seen as a serious threat to local long-term economic and social well-being. Building community by creating some sort of shared sense of well-being is seen as a way to overcome, or avoid creating, the tragedy.

The Problem of the NIMBY Syndrome

The NIMBY syndrome, another major contributor to unsustainability, is also thought to be a product of rampant individualism and self-interest. The NIMBY syndrome is a concept that describes the presence of classic free-riders in society, and is meant to capture the tendency for people to want certain benefits from business or government, and to not want to face the full costs of those benefits (Portney 1992). It is applied most directly to issues involving the siting of socially desirable facilities, including facilities that are designed to have net positive effects on the environment, such as hazardous waste treatment facilities. As such, it highlights the almost uniquely localized, site-specific, manifestations of free-ridership. The NIMBY syndrome says that people want various facilities, but they just do not want them in their neighborhoods or their communities. People want things made of plastics, but they do not want to live anywhere near a plastics manufacturing facility. They want electricity, but they do not want to live anywhere near an electric generating plant. They want toxic chemical wastes to be recycled, but they do not want to live near a facility designed to accomplish this goal. Of course, not all people want the goods, and not all people who want the goods oppose siting the facilities necessary to support them. The point is that in exercising their individual liberties, people are free to, and large numbers of people often do, shield themselves from facing the full costs of the goods and services they consume. Stated another way, rampant NIMBYism represents efforts of individuals to shift the environmental consequences of the benefits they receive to someone or someplace else.

NIMBY syndrome-based opposition to confronting environmental problems may be cloaked in a variety of different rationales. Frequently, people who oppose siting facilities in their communities base their opposition on some conception of fairness or justice (why should we have a new facility here when we already bear disproportionate burdens from other facilities?), or lack of trust (we can't trust the parties that build or operate the facility to do it safely), or other justifications. The bottom line, however, is that the NIMBY syndrome motivates the shifting of the ecological footprint to a geographic area that usually extends outside of the particular city.

To many commentators, the problem is that people want too much; there is too much consumption. And this may be true. But the NIMBY syndrome concept suggests that the real problem is the mismatch between what people consume and the prices they pay (not just the sticker prices, but the whole range of prices) for what they consume. Essentially, the NIMBY syndrome describes the fact that people are able, through the exercise of their rampant individualism, to reap the benefits and pay less than full price in doing so, or pay less than someone else. There is, of course, an inherent irony of individualism embedded in most NIMBY conflicts. The irony is that the conflict often consists of competing forms of rampant individualism as manifested in different views of private property rights, one bent on developing land in whatever way the owner wants regardless of the environmental impact, and the other bent on protecting property from the environmental consequences of that development. From an ecosystem and ecological footprint perspective, this means that people are often free to pursue ways of expanding their footprints, imposing costs on the larger ecosystem. When people want to consume plastics and do not want to live near the source of production, they force a broader ecological footprint. They force the manufacturing facility to locate elsewhere, in some other community or city, thus shifting and imposing their ecological impacts on other people or other places.

In perhaps another twist of irony, another viewpoint has emerged with respect to the effect of local opposition to facility siting and its effect on local sustainability. Whether local opposition contributes to sustainability or unsustainability appears to have much to do with what kind of facility is proposed to be sited, and what the alternatives to siting might be (Walsh, Warland, and Smith 1997). If the alternative to siting a hazardous waste treatment facility within a city is to export the hazardous wastes, then this would seem to provide an example of that city expanding its ecological footprint, not something generally considered consistent with attempting to become more sustainable. On the other hand, if the alternative is to force a reduction in the amount of locally produced hazardous waste, then it is difficult to avoid the conclusion that the net effect of the NIMBY syndrome on sustainability is positive.

Of course, there is no mechanism in place to ensure that people must bear the full price of their consumerism or other individualized choices. The core value of individualism not only undermines the efficacy of the belief that people should pay the full price; it also motivates people to seek this outcome through the political process. Without any explicit political or deliberative process to influence or coerce people to take an aggregate look at the consequences of their values, there is no incentive for them to do so. In theory, at least, elites (such as elected public officials) possess this responsibility, but in practice, elected officials too have little incentive to act on behalf of some community good, especially to the extent that the electorate cast their votes in concert with their individual value preferences. So, like the problem of the tragedy of the commons, the NIMBY syndrome makes the pursuit of sustainability a difficult political task. This has even given rise to the alternative acronym applied to elected public officials—NIMTOO—not in my term of office. In short, if the incentives for local elected officials are driven by the same individualism that directly creates the NIMBY syndrome, as one would expect it to be, then there is no public mechanism for overcoming the syndrome even when doing so is in the best interest of sustainability.

As with the tragedy of the commons problem, the NIMBY syndrome is often attributed to rampant individualism and the associated loss of concern for community. Solving the NIMBY problem accordingly seems to require attention to social and political mechanisms that facilitate or coerce people into facing up to the broader consequences of their seemingly contradictory social choices. Sometimes market-based solutions to NIMBY responses are advocated, where those who oppose facilities in their neighborhoods are economically compensated in some way. Frequently, however, solutions focus on this problem as one of governance, with prescriptions calling for building civil society, for efforts aimed at building community, and for promoting increased participation in public decision-making and governance. To many, the difficulty in these prescriptions is that they bump directly into common conceptions of the NIMBY syndrome in the first place. NIMBY responses, so this argument goes, are the result not of rampant individualism, per se, but rather of the exercise of this individualism in the context of localized issues. In this view, promoting participatory processes, and encouraging more interpersonal kinds of participation, invite even greater parochialism of the sort that stands at the heart of NIMBYism. According to this view, because the pursuit of sustainable cities requires higher levels of cooperation than usually achieved, greater participation will make pursuit of sustainability more difficult, not less difficult. Whether there is evidence concerning the possibility that participatory governance processes can overcome rampant parochialism and individualism is perhaps an empirical question, one that is addressed later in this chapter.

Transboundary Environmental Impacts

The third generalizable social and governance problem that impedes the pursuit of local sustainability is the creation of, indeed the reliance on, transboundary environmental impacts. If the NIMBY syndrome is a mechanism through which people in a city are able to avoid paying the full price of the goods and services they consume, one consequence is the shifting of costs across geographic boundaries. Transboundary pollution issues in particular are often discussed in the context of problems that arise between or among nations (Ingram, Milich, and Varady 1994; Pallemaerts 1988; Young 1994). But they also characterize environmental problems between or among governmental entities within a country, including intercity and city-suburb jurisdictions. Transboundary environmental impacts are effects of activity within cities that transport environmental impacts to places and people outside of the city, or effects of activity that occur outside the city that impact the people inside the city. Indeed, the extent of the transboundary problem helps to define the size and scope of any city's ecological footprint, where cities with larger footprints presumably exhibit greater transboundary impacts. To the extent that a city's ecological footprint extends beyond its own jurisdictional limits, this may be said to constitute a transboundary problem.

The presence and prevalence of transboundary impacts constitute primary rationales for elevating environmental solutions to larger geographic areas, particularly in an effort for one city to solve the environmental problems that are imported from other places. For example, one of the reasons why some are skeptical of the efficacy of pursuing local sustainability is the contention that transboundary problems cannot be solved at the local level. In order to address air pollution, for instance, a broad geographical perspective is needed. The City of Hartford, Connecticut, frequently experiences significant problems with air pollution, only a portion of which is due to the activities that actually take place within Hartford. Instead, the air quality problems in one city are often, at least in part, created by pollution emitted somewhere else, frequently in other states or even in other nations. Much the same problem exists in most large cities, particularly in the Eastern and Midwestern states. There are many other environmental and pollution problems in cities that are analogous to this air pollution example.

Common wisdom suggests that cities have no incentive to deal with the ecological impacts that are created within their jurisdictional boundaries, particularly if market forces make it relatively easy and cheap to export those impacts. As was the case with the NIMBY syndrome, rampant individualism contributes to a political result in which local leaders find no incentive to confront the environmental impacts produced in their cities. Indeed, the primary problem with the NIMBY syndrome is that it influences, perhaps even makes highly likely, transboundary exportation of environmental impacts. Thus, so the argument goes, in order to overcome such problems, policies must be pursued at a level that takes into account both the interests of those who export environmental impacts and those who experience the consequences of those impacts. Accordingly, so the argument goes, policies must be pitched at this higher level because it is only here that there is self-interest motivation for a government agency to coerce one city government into reducing its impacts on others (Ophuls and Boyan 1992, 195–206). Thus, command and control government regulation imposed by one level of government on another is thought by many to constitute the only real solution to transboundary environmental impacts. Yet, as developed next, communitarians frequently argue that there is a plausible alternative solution.

Communitarian Solutions to the Three Problems?

It is relatively easy to describe the three generalizable social and governance problems that are thought to play such a pivotal role in impeding the achievement of local sustainability. It is quite another to prescribe ways of overcoming these problems. Virtually every effort to prescribe solutions to the three problems focuses on some sort of effort to engage the public, to rebuild a sense of community, to rebuild civil society, or to otherwise empower people to make public decisions and to confront the often difficult trade-offs these decisions entail. Not every city needs to overcome all three of these problems, and indeed, one of the lessons from the work of Clarence Stone (1993) is that some cities seem almost predisposed to be what he calls "middle-class progressive cities." In these cities, basic public support for environmental protection can be taken for granted, and the level of participation and cooperation among seemingly disparate interests ensures the marshalling of the necessary resources to achieve environmental (and other) policy goals. Yet the vast majority of cities are not like this, and these problems do represent major political issues that communitarian approaches seek to address.

As easy as it is to advocate greater public participation as a potential solution to the three impediments to achieving sustainable cities, it is far more difficult to articulate the specific reasons why community building may be expected to produce more sustainable places, and the form that participation should take to accomplish this. Yet this is precisely what communitarian thinkers and many sustainability advocates attempt to do. Essentially, most communitarian prescriptions look for ways to build or rebuild particular elements of American civil society with an eye toward generating "community." What this means is that such prescriptions propose creating social and political institutions and associated processes that they believe will have the consequence of altering the ways people behave, especially in how a city's residents interact with each other. Perhaps even more important, there is an expectation that these institutions and processes will play a transformative role in forming or changing residents' values. In other words, these institutions' primary functions would be to produce residents who understand and come to terms with a fuller range of costs their consumer and environmentally impacting behaviors impose on others, and as a result, their values and behaviors would be altered. Although participation in organizations and in public policy decision processes can be advocated from many theoretical bases other than those that are communitarian, clearly for many cities, participation and civic engagement constitute means to what is largely a communitarian end. The end for some appears to be changing attitudes, values, and behaviors. For others, it is explicitly the building of community and empowering the residents of the city. And for still others, it is motivated by Ostrom's findings that providing opportunities for individuals to communicate may foster creation of rules of governance that produce optimal environmental outcomes. Either way, communitarian approaches provide comfortable theoretical justifications or rationales. But how is this supposed to be accomplished? How is it that engagement and participation might be expected to produce these communitarian results?

Most communitarian prescriptions for rebuilding civil society in ways that facilitate the requisite transformations rely on engaging people in organizations and institutions, organizational processes, and decision processes. Indeed, virtually all of the prescriptions from the National Commission on Civic Renewal were grounded in the notion that engaging residents in various kinds of civic and political organizations hold the best hope for affecting peoples' attitudes. Some examples of participatory democracy will help to clarify this line of reasoning. Most of these examples come from the "deliberative democracy" literature, studies that examine the role of participation in strengthening public policies and requisite attitudes toward government. They are not examples of how *sustainable cities initiatives* have successfully engaged people. Rather, some are general examples that illustrate the reasons why some have suggested that civic engagement represents an important element in any effort to build or rebuild community. Others are examples of how participatory processes have been employed to address a variety of local environmental issues.

Is There Empirical Support for Communitarian Solutions?

Skeptics of the communitarian solutions note that despite the logic of their approach, it simply does not work in practice. In today's era of high technology, where the average person has neither the technical/scientific background nor the skills to fully comprehend the complexities of what might be necessary to create a sustainable community, and certainly does not have the time to engage in highly participatory decision processes, it is unreasonable to expect a city's residents to become part of a communitarian solution. Small-scale public participation in relatively narrowly defined and technically unsophisticated issues may be possible, although not easy, and the results of that participation are rarely predictable in terms of results. Beyond that, it is unreasonable to expect public participation to transform governance or to contribute to creating a sustainable city in any meaningful way. Or so the argument goes.

Yet in recent years, the communitarian prescriptions have been subjected to a variety of empirical analyses, and have shown significant promise. Efforts to engage residents in dialogues over issues of planning, policy directions, and specific programmatic problems have sometimes demonstrated their effectiveness. Locally organized public involvement efforts, whether focused on local, statewide, or national issues, not only provide mechanisms for residents to have their voices heard and for government to respond on specific issues, they also provide mechanisms of civil society that actually change the people who participate in them. The experiences of those efforts produce something of a mixed bag of results—far less compelling than most communitarians or advocates of sustainability would hope for.

One of the more impressive efforts at creating deliberative democracy in action came as the result of the "Americans Discuss Social Security" initiative spearheaded by the Pew Charitable Trusts during 1998 (Urahn and Ledger 2000). A large portion of this initiative focused on creating and hosting local nonpartisan forums in five major U.S. cities where residents were invited to participate in extensive discussions about what should be done with Social Security. The importance of this is twofold. First, it provides a sense of whether deliberative processes are capable of helping to overcome narrow self-interest, since the Social Security problem resembles a common-pool resource issue. Second, it provides an opportunity to assess whether deliberative policy decision processes can be sustained over time.

Several hundred residents were brought together in each of the five cities, Austin, Buffalo, Seattle, Des Moines, and Phoenix. Not only was this initiative treated as an opportunity to engage the public, it was also designed so that a serious effort could be made to assess the consequences of participation. The forums were assessed on a variety of criteria. Although no explicit effort was made to directly examine the "community building" effects of the discussions, there was an effort to assess whether participants in the discussions were more likely than random samples of nonparticipants to come to a consensus about preferred policy changes, and report an interest in future participatory activities. Did consensus develop? Did participation in these discussions energize people to participate in other activities, including opportunities to influence policy decisions on Social Security? Apparently neither of these results occurred, at least not to the degree expected. Participants were no more likely than others to stay committed to and engaged in discussions about the future of Social Security (Cook and Jacobs undated, 23–24; Cook, Barabas, and Jacobs 1999, 20–21). At least on its face, these results would suggest that to the extent that fostering continued interpersonal interactions represents community building, the participatory practices from this initiative did not seem to make a difference. Yet, it is also true that most of these cities might already be said to be highly participatory places. Civic engagement and participation in neighborhood associations is unusually high in Seattle, Buffalo, and Austin (Berry, Portney, and Thomson 1993).

More directly related to environmental issues, a number of studies have focused on the role of participatory processes in siting hazardous waste facilities of various sorts. Such studies have specifically examined the role of participation in aiding or helping to overcome the NIMBY syndrome. Does greater participations increase NIMBY responses, or does it help alleviate them? In general, most of the case study analyses seem to offer some support for the idea that engaging affected parties and residents in deliberative decision processes do indeed help to overcome NIMBY reactions. For example, Rabe's study of hazardous waste facility siting in Alberta, Canada, points to the need for a long-term commitment to engaging the public, and to being responsive to the detailed objections of affected parties (Rabe 1992). In a comparative analysis, Williams and Matheny (1995) suggest that carefully structured democratic dialogue in the context of hazardous waste facility siting is capable of yielding collective decisions that transcend parochial responses. And Mazmanian et al. (1988) have argued that creating dialogue among parties with long-standing differences helped to break the gridlock on hazardous waste policies in California.

Another example of the way that participatory processes are thought to contribute to overcome mainstream NIMBY issues comes from DeWitt John's work advocating civic environmentalism. John suggests that in Seattle and its surrounding King County during the 1980s, widespread participation transformed the public decision-making process (John 1994, 10–14). When public officials wanted to respond to the closing of major landfills used to dispose of the city's solid waste by building a number of incinerators, the residents reacted in predictable NIMBY fashion. But instead of simply blocking the siting and construction of incinerators, the participatory process stimulated the development of a comprehensive recycling effort to eliminate the need for either landfilling or incineration of solid waste. Presumably, the end result was exactly what civic environmental advocates seek—a solution that was better for the environment, and more sustainable, than what would have resulted in the absence of participatory processes.

Communitarian Elements of Sustainable Cities

While communitarian solutions to the serious issues that tend to help undermine pursuit of local sustainability may seem at least somewhat plausible in theory, and may even find some indirect support in a variety of specific settings, to what extent can communitarian practices be said to be part of actual sustainable cities initiatives? In the best of all research worlds, it would be desirable to be able to point to analysis of the extent to which sustainable communities initiatives' participation processes can be said to have altered peoples' values, or at least started the process of rebuilding community. Unfortunately, there are no such analyses to point to. There are, however, some other critical questions about sustainable cities initiatives' communitarian foundations that reveal much about those initiatives. Are communitarian elements—elements that stress institutions and processes in civil society that are concerned with community building and interpersonal interactions of various sorts—integral to specific sustainable cities initiatives? To what extent, and in what ways, have specific cities' initiatives been grounded in some conception of community and community building? What are these communitarian elements, and how are they manifested in practice? These questions may be addressed by exploring the manifestation of communitarian elements as they have found their way into the operation of cities' sustainable communities.

To state it succinctly, the central focus of communitarian elements of sustainable cities initiatives is the nature of the participatory processes in the cities. These participatory processes relate to two different aspects of sustainable cities. First, how participatory are the processes used to develop cities sustainability plans, particularly the processes associated with indicators projects? As noted in chapter 2, many indicators projects appear to be based on the notion that the indicators project itself can and should serve as a primary vehicle for engaging the residents of the city in community-building exercises. Second, cities may elect to include "participation" indicators as explicit measures of sustainability. When a participation indicator is included in a sustainability plan, this represents a clear statement that community building ought to be a goal.

Even in cities where high levels of participation in political and governmental decision processes are not evident, it is still possible for cities' sustainability initiatives to seek to "build community." For the most part, building community refers to efforts aimed at promoting greater interpersonal interaction, greater participation in civic organizations, and, in short, fostering civil society. For example, Seattle's Comprehensive Plan 2002 lays out a broad array of community building goals and policies as part of the Plan's human development element. These goals and policies include seeking to "make Seattle a place where people are involved in community and neighborhood life...", "work toward achieving a sense of belonging among all Seattle residents...", and "promote volunteerism and community service..." (City of Seattle 2002). The key, however, is that the plan calls for a variety of specific city actions to promote this community building. As presented in chapter 2, among the twenty-four cities examined here, twenty-one have sustainability initiatives that promote some level and type of community building through greater participation. Of course, not all of these efforts are the same.

In Santa Barbara, California, an effort is made to promote "civic engagement," and the South Coast Community Indicators Project specifies three indicators of civic engagement (Santa Barbara 1998). The first is the total dollars collected by the Santa Barbara United Way; the second is the turnout in local elections; and the third is ticket sales from performing arts. Although these indicators resulted from a process that included eleven public meetings, this project does not appear to make provision for any sort of ongoing participatory processes. Moreover, the indicators project itself is not accompanied by any sort of action plan, specification of programs or policies that should be pursued to promote more civic participation, or interventions that should be taken if the indicators point to declines over time.

A third example comes from the city of Jacksonville, Florida, where the Quality of Life Indicators Project includes an effort to monitor the amount of community participation, and measures the amount of community participation annually through its telephone-based poll. Adult residents are asked whether they have given their time, without pay, to any charitable, religious, or volunteer organizations. The project established a target of increasing reports of community participation to 75% of the adult population, an increase from the reported 60 to 65% found in recent surveys (Jacksonville Community Council, Inc. 2000). However, there does not seem to be any significant programmatic mechanism in place for actually promoting more volunteerism.

What these three examples have in common is that they all specify community building as an important part of building a sustainable city. It is perhaps surprising that, given the decision to include community building, there is such little explanation for what the cities expect to achieve by building community as they define it. While communitarian concepts often seek community-based participation as a mechanism for creating neighborhood stability, and for contributing to overcoming the three deadly sins associated with being unsustainable, it is certainly not clear that the designated community-building activities and indicators bear any particular relationship to these goals. If a city achieves its target of increasing volunteerism, will that city necessarily achieve anything else? If Jacksonville increases its reporting of volunteerism to 75% of the adult population, will Jacksonville experience fewer NIMBY responses? Will it experience fewer situations involving the tragedy of the commons? The same questions can be posed for the community-building or civic engagement goals and indicators in most cities. In short, it is not clear what the relationship is between cities' community building efforts and the achievement of other goals that are more directly related to sustainability.

Another way of looking at the issue of participation as community building is to examine whether cities sustainability initiatives themselves have been based on some form of participatory process. Short-term participatory processes that may be associated with the specific task of developing a sustainability plan or an indicators project are but weak substitutes for long-term, open-ended participatory processes. But in practice, they still provide some insight into community-building opportunities made possible by cities' sustainability efforts. As Kline notes, one of the sometimes unstated goals of sustainable cities projects is to create a collaborative community participation process that will result in greater democracy and community-building (Zachary 1995, 7).

The process of developing sustainability initiatives in most cities does not seem to rely on any sort of extensive participation activities. As a general rule, sustainability initiatives seem to prefer polling and survey research techniques to obtain "resident input" into the process. Of course, such polling and survey research can, if done properly, provide much more accurate information representative of the entire city population than participatory processes. Reliance on such techniques does not, however, promise to directly be part of any community building process. In other words, if the goal of engaging residents in the sustainability initiative is to make some contribution, however small, to building community, then most cities do not do this.

There are, however, examples of processes that do seem to take seriously the idea that the development of sustainable cities programs itself can be a part of the community building process. Either in the context of the development of a sustainability plan itself, including the development of sustainable indicators, or in the context of a broader comprehensive planning process, many cities use some form of "visioning" process to engage residents. The primary purpose of such visioning processes is to get input from interested residents concerning the directions in which they would like to see the city move. Presumably, this purpose includes making the city government more responsive to the wishes of the populace, or at least an active part of it. Rarely are such visioning processes thought of as mechanisms for community building or for changing the way participants think about sustainability, but that may in fact be a result.

The two most frequently cited examples of indicators projects where participatory processes were integral are Seattle, Washington, and Cambridge, Massachusetts (Zachary 1995, 27–29). In the nonprofit based Sustainable Seattle project, the process of developing the sustainability indicators involved participation by over 250 volunteers who engaged in a series of civic forums and smaller committee group meetings. AtKisson (1999) reports that these participants were elites who were self-selected after about 300 "people in positions of responsibility in government, business and a wide variety of civic organizations" were invited to participate in the process of developing indicators of sustainability. The apparent purpose of this involvement was to maximize the likelihood that a broad consensus would emerge in support of the use of the developed indicators. There was little explicit attention given to the possibility that the process of involving these volunteers would itself constitute a part of a community-building effort, but there are clear undercurrents to that effect. A number of the indicators that were developed sought to measure the extent of community participation in city activities. But Seattle's initiative embraces a commitment to participatory and interpersonal interactive processes well beyond that found it the indicators of sustainability. In the city government's 2000 draft comprehensive plan "Toward a Sustainable Seattle," there is an entire "element" devoted to "human development," where "building supportive relation-ships within families, neighborhoods, and communities" is an explicit goal. Equally important is the fact that this is accompanied by a delin-eation of actions and programs that the city intends to engage in to promote greater participation in planning processes. It is perhaps less surprising that Seattle would define sustainability to include community-building goals and activities than it is that there would be so little justi-fication or explanation for doing so.

As part of its sustainable indicators project in Cambridge, Massachu-setts, the Cambridge Civic Forum developed a participatory process generally referred to as "reaching in," where volunteers from the Forum would attend meetings of other civic organizations and community groups to elicit their input and visions for sustainability. As with Seattle, there was no stated goal of using the participatory processes associated with the indicators projects, or sustainability initiatives in general, to contribute to democracy or to community building. Yet the participatory processes associated with both of these efforts seem to recognize that if community building is an important target of sustainability indicators, then the development of those indicators must be done with as much participation as possible. Given the limited participation reported in the scattered descriptions of these projects, and the fact that most of that participation emanates from special interest "stakeholder" groups, one can only speculate about its impact on community building.

The Politics of Becoming a More Communitarian Place

Inclusion of such communitarian elements may seem to take sustain-ability far beyond what some envision—particularly for those whose vision is focused mainly on the physical environment. Indeed, if taking sustainability seriously in cities requires that they become more com-munitarian in their orientation, requires that they develop a healthy dose of communitarian values, this is a tall order to fill. Most cities are not communitarian kinds of places, and any local initiative that requires them to become so in order to take sustainability seriously may be doomed to failure. Yet, a number of cities' sustainability initiatives seem to have been founded on the notion that pursuing sustainability can and should serve as the vehicle for ushering in social values that are heavily communitarian. In short, the politics of sustainability in cities is very often about the politics of trying to transform cities into more communitarian kinds of places, and the struggle involves the clash between those who advocate this approach and those who represent and support the status quo.

Sustainability initiatives, particularly those that are heavily based in the nonprofit sector, often experience substantial frustration from the fact that their efforts may not produce the kinds of communitarian results they thought would accrue. When city officials, such as the mayor, city councilors, and agency directors respond in lukewarm fashion to the prospect of taking sustainability seriously, sustainability advocates frequently must confront the political reality that their initiatives turn out to be a less than effective mechanism for creating community, empowering people, or producing the social changes they sought. For example, in Boston, the mayor has played a critical role in determining how far the sustainability initiative has gone. In response to a question concerning the future of sustainability's potential for enhancing a sense of community in that city, a representative of the Boston Foundation, the organization that spearheaded the initial effort, questioned whether the mayor of the city would be willing to support sustainability in the face of opposition and skepticism. She noted that: "Well, it's no secret that if Mayor Menino doesn't make it a priority and doesn't embrace the project, it will not work . . . he obviously will have the greatest impact on how effective the project is and what kind of time frame we're looking at" (Boston 2000). In Olympia, Washington, the Sustainability Roundtable found that the city was more interested in its indicators project for symbolic rather than more substantive reasons. A leader in the Roundtable noted that initial support for the indicators project ". . . certainly wasn't for any actual concern for the environment or our future. In 1995, Olympia adopted 'sustainability' as its guiding principle—for political reasons mostly—and it needed something to show for its new principle . . . [but] . . . as for the real political power, though, for the most part, they have better things to worry about. . . . Soon after the original indicators project was completed, the city no longer wanted an affiliation with [the Roundtable]. In fact, sustainability in general seems to be 'out'" in Olympia (Olympia 2000).

To date, it is difficult to find cases where there is evidence that sustainability initiatives' participatory processes have successfully transformed the values of city officials or residents. Although participatory processes associated with sustainability initiatives may well have helped to crystallize commitment to the environment, they seem unable to overcome true political opposition that might be present. Indeed, the cities that seem to take sustainability most seriously would appear to be cities where little political opposition existed in the first place. Clearly, initiatives that anticipate serious political opposition tailor their approaches to avoid confronting that opposition. That may well be the reason why so much of the activity in Chattanooga and Jacksonville, to name two such places, is so concentrated in the nonprofit sector. It may well be that keeping sustainability issues out of the public sector helps to avoid confrontations with interests that might be opposed to sustainability for fear that sustainability would end up the loser. In other words, as a matter of initial strategy, pursuing sustainability through the nonprofit sector is a very different, perhaps easier, challenge than pursuing it through city government. Nonprofits that seek to create and enhance a sense of community would seem to be less likely to encounter political opposition than government agencies trying to accomplish the same goals. In this sense, San Franscisco, whose sustainability experiences are discussed in more detail in chapter 7, may serve as an example of what happens when the community-building initiative confronts serious opposition. Suffice it to say that in most cities with sustainability initiatives, trying to build community adds a level of political difficulty to a task already facing significant challenges.

Communitarianism and Sustainable Cities Initiatives: A Summary

If the purpose of becoming more participatory is to promote the fundamental underpinnings of sustainability, then most cities that purport to pursue sustainability appear to come up short. This conclusion is supported by at least three observations from a wide array of cities. First, there is little recognition of the relationship between the presence of three deadly sins and the inability to become sustainable. While the more theoretical treatments of sustainable communities suggests that the tragedy of the commons, the NIMBY syndrome, and transboundary shifting represent significant impediments to achieving sustainability, at least at the conceptual level, there is no apparent recognition of this link in most cities' sustainability initiatives. Second, there is no recognition that participatory processes, as conceived by communitarians, hold promise for overcoming these three deadly sins. Communitarians propose that enhanced community building and the interpersonal sense of connectedness that come from civic participation are capable of helping to undermine the three deadly sins. Accordingly, if there is to be progress toward achieving sustainable communities, there needs to be a great deal of attention to fostering community-building processes. This has certainly happened in some cities, and is sometimes part of such cities' indicators of sustainability. But in most cities, these issues are never really addressed. And third, even in cities that have built civic participation into their sustainable communities initiatives, the efforts appear so limited as to make unlikely substantial progress toward overcoming the three deadly sins. Of course, it remains to be seen whether those cities, such as Seattle and Cambridge, are able to actually make more progress toward becoming sustainable than cities that have not addressed community building in their sustainability efforts.

6

Is a Sustainable City a More Egalitarian Place? Sustainable Communities, Environmental Equity, and Social Justice

Sustainable cities initiatives in the United States and elsewhere sometimes elect to address issues of "justice" or "equity" in their activities. Such activities reflect concerns about general social justice, and sometimes they reflect more explicit concerns specifically about environmental justice. Social justice concerns tend to focus on the distribution or maldistribution of income and wealth, housing quality, employment opportunities, crime, health and well being, and access to a whole range of public services. Environmental equity or justice, often thought of as a subset or special type of *social* justice, involves issues where the focus is specifically on the distribution or maldistribution of environmental consequences, which usually translates into concerns over inequality in exposure to environmental hazards and risks. The purpose of this chapter is to review some of the extant literature on social and environmental justice to highlight the extent to which social and environmental justice might be said to be linked to sustainability in general, and by extension, to sustainable cities. This chapter does not break new conceptual ground on these issues. Rather, after reviewing existing conceptual discussions, this chapter examines how some specific sustainable cities initiatives have internalized issues of social and environmental justice, and will contrast these cities with other sustainable cities that have not.

Over the past twenty years or so, there has been considerable interest in issues of social justice, particularly with regard to the third world. Probably since the work of the Brundtland Commission in the 1980s, this concern about social justice has become linked to issues of sustainability. The Brundtland Commission clearly articulated the beliefs of many when it argued that sustainability requires greater social justice and economic equality. This, as seen shortly, is not an undisputed conceptual claim, and the connection between equity and sustainability is not always so readily accepted in cities that otherwise seem to take sustainability seriously.

At the same time, there has been a virtual explosion of interest in issues of environmental justice. Environmental justice is largely concerned with the ways that decisions made mainly in the public and private sectors of the American economy intentionally or unintentionally impose differential environmental risks on particular groups of people. Perhaps beginning in the early 1980s, an increased awareness of these differential effects, believed to primarily fall on people of particular racial or ethnic minority backgrounds and the poor, fueled the "environmental justice movement" in the United States. Although the environmental justice movement is national and international in scope, most of its targets are local. Virtually all of the specific manifestations of environmental injustice or "ecoracism," to use one of Robert Bullard's provocative terms, occur at the local level.

As awareness of actual and potential differential effects increased, concerns about such effects found their way into the discussions of many sustainable cities initiatives. Primarily reflected in elements of some cities' indicators of sustainability, explicit efforts have been made in some places to minimize these differential effects. At least in part as an extension of the idea of sustainability beyond purely environmental issues and into "quality of life" concerns, issues of environmental justice have broadened the scope of what sustainability means in the context of cities. Not all cities pursuing sustainability seem to share a concern for such issues. In other words, many cities that purport to be working toward becoming more sustainable do not address issues of equity at all. Indeed, as this chapter explicates, whether some form of social or environmental justice is part of cities' sustainability initiatives certainly constitutes a major differentiating characteristic. The purpose of this chapter is to take a broad look at the relationship between local social environmental justice issues and the pursuit of sustainability in cities, and to examine the ways that some cities have incorporated issues of environmental justice into their initiatives. It should be made clear from the outset that this analysis does not purport to assess how effective cities' efforts at addressing social justice and environmental equity are. Rather, this analysis is designed to try to understand how cities attempt to address these issues within the context of their sustainability initiatives. Before examining the ways that cities differ in whether and how they treat issues of social and environmental justice, a general discussion of social and environmental justice issues is warranted.

Sustainability and Social Justice

Although, as is discussed later in this chapter, many cities sustainability initiatives incorporate social and economic equity concerns into their efforts, it is not altogether clear why equity and sustainability are sometimes linked. Moreover, some cities that otherwise appear to take sustainability very seriously never mention or address issues of social or economic equity at all. This certainly raises the conceptual question as to what the connections between sustainability and social justice are. Stated succinctly, does a city have to address social justice issues in order to be said to take sustainability seriously? Or stated another way, can a city essentially ignore social and economic equity issues and still take sustainability seriously? Part of the answer to this question lies in how one defines sustainability in the first place. This is a large and complex conceptual issue, one that cannot be fully considered here. A brief discussion of the connection, or lack thereof, between equity and sustainability helps to explicate this issue, and provides a foundation for the discussion of cities' attempts to address equity in their sustainability efforts. Suffice it to say here that whether there is a link between social justice and sustainability is a matter of debate and definition, and that the reason why this is important is because some cities seem far more comfortable than others in making the assumption that there is an "inextricable link."

In the broadest conceptual sense, advocates of sustainability frequently argue that the pursuit of sustainability and sustainable development are inextricably linked. Perhaps most clearly articulated in the Brundtland Commission report, it is thought "inequality is the planet's main 'environmental' problem" (WCED 1987, 6). But there are few in-depth discussions of whether there really is such a link, and there are even fewer empirical analyses that provide evidence for its existence. The central issue is whether achievement of greater social justice and sustainability are dependent on, or independent of, each other. From a conceptual perspective, Dobson states it succinctly in his analysis that aims

... to assess the theoretical relationship between environmental sustainability and social justice ... [because] ... we cannot assume that these objectives are compatible, and their potential incompatibility raises issues of political legitimacy for them both. ... it is just possible that a society would be prepared to sanction the buying of environmental sustainability at the cost of declining social justice, as it is also possible that it would be prepared to sanction increasing social justice at the cost of a deteriorating environment. (Dobson 1998, 3)

Dobson admits that societies would not likely choose to pursue one goal to the exclusion of the other, but his point is that achievement of these goals is largely independent. It may well be possible to achieve one without the other.

Much of the argument concerning the connection between the two derives from the logic of the Brundtland Commission report which argues that "... poor people are forced to overuse environmental resources to survive from day to day, and their impoverishment of the environment further impoverishes them, making their survival ever more difficult and uncertain. ... those who are poor and hungry will often destroy their immediate environment in order to survive" (WCED 1987, 27–28). The problem with this logic for Dobson is that

Prima facie, it is unlikely to be true that poor people are always forced to overuse environmental resources, since 'overuse' implies an already existing scarcity. Poor people do not always and everywhere live in conditions characterized by environmental resource scarcity, so the conclusion reached by the Commission is not as universally relevant to environmental sustainability as its report suggests. ... poor people are often, of necessity, absolutely aware of resource problems, and have developed successful and sustainable strategies to cope with them. This suggests that careful analysis of exactly where and why poverty induces unsustainable overuse of resources is required. (Dobson 1998, 15–16).

In some places, there is reason to believe that there is a relationship between poverty and the geographic distribution of environmental problems. McGranahan, Singsore, and Kjellen argue that the environmental threat from social inequity and poverty derive from particular spatial distributions of environmental impacts. In an analysis of the cities of Accra, Ghana, Jakarta, Indonesia, and Sao Paulo, Brazil, they suggest that:

Whether one looks across cities, at the history of affluent cities, or even across different groups within a city, it is possible to discern the outlines of an environmental transition relating to affluence. As affluence increases, there tends to be a spatial shifting of environmental problems from the local to the regional and global. There also tends to be a temporal shifting from immediate health problems to, in the case of global warming, intergenerational impacts. These tendencies only reflect dispositions: policies, as well as demography and geography, can make an enormous difference, and vastly different environmental conditions can be found both within and among cities with comparable affluence. Generally, however, the poor create environmental problems for themselves and their neighbors, while the affluent distribute their environmental burdens over an expanding public. (McGranahan, Songsore, and Kjellen 1996, 105)

In the context of American cities, the connection may seem fairly tenuous particularly because the kind of poverty that seems to favor the creation of local environmental problems is not prevalent in the United States. In what ways might a person living in inner-city poverty be said to "overuse" environmental resources? Is it possible that such a person would actually have a smaller personal ecological footprint than a middle-class person whose disposable income permits greater consumption of consumer goods? To the extent that sustainability relates to the environmental impacts produced by residents of a city in their various behaviors, this raises questions about whether greater equity would produce greater sustainability. All of this is to say that it is not entirely clear that it is necessary for a city to pursue social justice goals to take sustainability seriously.

Although many advocates of sustainability seem to be willing to accept the link as inextricable, few provide much in the way of explanation for why this link is thought to exist. For many, the idea that social justice must be pursued as a component of sustainability is an assumption, or starting point, that needs no explanation. This, of course, comes from the idea that the definition of sustainability is, by its very nature, value-based, or what Robinson et al. (1990) refer to as a "normative ethical principle." As such, if one places high value on achieving social equity, if one considers high degrees of social equity a desirable characteristic for society, then of course it must also be a desirable characteristic for a sustainable society. For Kline the connection between equity and sustainability is also an assumption. Incorporating equity issues into indicators of what she calls "disparities," which represent a dimension of "economic security," she notes:

It is important to evaluate disparities in income, in lending, in dollars that remain in a community compared to those that leave, in employment distribution among employees, and in how dollars are spent. The range of items measured and the trends over time tell a lot about a community's long term stability and the fluctuations it faces over time. For example, a trend of out-migration among young adults and a concentration of young children and older people affect the economic base of a community. *An assumption is made that the more diversity, in general (not necessarily in every area), the more sustainable a community is likely to be.* [emphasis mine] (Kline 1995a, 15)

The problem is that if social and economic equity issues are linked to sustainability merely by assumption or as a social construct, then it may be just as reasonable to assume that equity is not an explicit or an important element of creating a sustainable city. Indeed, many cities that seem to be eager to pursue sustainability do not seem compelled to deal directly with issues of social equity, or to tackle the potential political implications that underlie them. The central question is: why does there need to be equity in order to produce sustainability? Unless there is a compelling logic to support this connection, many cities will see little reason to incorporate equity considerations into their sustainability efforts.

Environmental Justice: A National Movement and Its Local Manifestations

Although this is not the place for a detailed account, the environmental justice or equity movement in the United States has grown out of a special concern for the uneven ways in which environmental risks appear to be distributed among or across people. Although there are at least four different dimensions to issues of environmental equity—the distribution of environmental risks across social groups, across geographic areas, across generations (over time), and across species—the term is most commonly associated with unequal risks across social groups (Haughton 1999). Environmental justice, in its most common form, specifically focuses on taking actions to reduce the environmental risks borne by people of minority racial or ethnic status and people from lower socioeconomic groups, usually taking as a given that such people are in fact discriminated against by being asked to bear disproportionately greater environmental risks. Whether it is in the place of residence, where there may be proximity to household toxins such as lead paint, asbestos, and industrial pollution and its consequences, or in the workplace, where employees may experience exposure to industrial toxics and hazards, minority and lower socioeconomic status people appear to be at greater health risk than most other people. Whether it is because of exposure to hazardous materials from an abandoned Superfund site, or to the toxic air emissions from a nearby plastics manufacturing plant, or to PCBs in a local water supply, or to the presence of lead paint in run-down housing, to name just a few of the many environmental risks that have been examined, there is deep concern that traditionally disadvantaged groups in American society bear much greater environmental burdens than others.

The environmental justice movement probably got its start in 1982 when a local chapter of the NAACP worked to organize people in Warren County, North Carolina, to oppose the siting of a new solid waste incinerator (Geiser and Waneck 1994; Rees 1992, 15–16). The incinerator threatened to expose the predominantly Black residents of the county to a variety of airborne contaminants, including PCBs and dioxin. But it was not until the publication of a report analyzing the siting of hazardous waste landfills by the U.S. Governmental Accounting Office (1983) that systematic evidence of racial discrimination began to surface. This report indicated that African Americans composed the majority population in three out of four communities of the southeastern United States where hazardous waste landfills were located. Indeed, much of the debate about environmental justice has focused on the issue of whether and to what extent *local decisions to site environmentally impacting industrial facilities*, including hazardous waste treatment facilities, solid waste incinerators and landfills, and many other types of facilities, are equitable (Portney 1991a; Walsh, Warland, and Smith 1997).

The movement received a major push from the 1987 work of the United Church of Christ (UCC 1987), which sponsored a study and analysis of the U.S. General Accountry Office (USGAO 1983) findings by Charles Lee focusing on the existence of hazardous waste sites in or near minority and poor communities. Dr. Benjamin Chavis, Director of the UCC's Commission for Racial Justice, interpreted the study's findings as reflecting "environmental racism." The movement received another major push with the 1990 publication of Robert Bullard's *Dumping in Dixie* (1990), which presented evidence to support the argument that the African American population is under systematic attack from environmental risks imposed on them by industry. To Bullard, it was no accident that all five of Houston's landfills and six of its eight incinerators were sited in African American neighborhoods. To him it is a clear example of "ecoracism." Since that time, numerous other studies have highlighted and tried to establish the nature of the racial and class basis for environmental risks (Bryant and Mohai 1990, 1992; Camacho 1998; Foreman 1998; Mohai and Bryant 1992a, 1992b; Perkins 1992; Walker and Traynor 1992).

The fundamental character underpinning environmental justice advocacy is the idea that our economy functions to the detriment of minorities and the poor because it forces them to bear disproportionate environmental risks. Some see this as an inevitable result of various kinds of failures of markets (Dore and Mount 1999; Goldman 1993), but many others see this disproportionate burden as an intentional result. Here, the blame is placed squarely on the shoulders of business and industry, whose practices in pursuit of economic growth and development have been termed "garbage imperialism" and "radioactive colonialism." As Bullard (1990, 1994) states in reference to the so-called "cancer alley" of Louisiana, "the plantation and slave economy in the rural parishes was replaced with the petrochemical industry as the 'master' and 'overseer'."

Despite the research making the case for ecoracism, until recently it has not always been totally clear that African Americans and other minority populations are necessarily at greater environmental risk, or what the magnitude of the differential risks are. Without going into great detail, even the UCC's study suggested that on a national scale, there is not a great deal of variation in the likelihood that people of a given racial or ethnic group would live near hazardous waste sites. According to the UCC report, nearly half of the population of the U.S. lives in communities with toxic waste sites (USEPA 1992, 16). The U.S. Environmental Protection Agency suggested in 1992 that, with the very notable exception of exposure to lead, there is a real lack of information concerning actual exposure to environmental risks necessary to be able to draw definitive conclusions (Anderton et al. 1994). Since that time, evidence has started to mount that, indeed, minorities and lower socioeconomic status people do bear considerably greater risk from exposure to environmental contaminants than other people. There is little question that the environmental justice movement has made significant strides in elevating the scrutiny that equity issues have received, in sensitizing people to biases in the way environmental risks are borne, and in beginning a process of creating checks against private sector abuses. What is less clear is the potential impact of achieving greater equity on pursuit of sustainability. In other words, when the primary focus is on achieving greater social or environmental equity, what does this necessarily mean for working toward sustainability?

One of the principle ways that environmental justice issues affect the pursuit of sustainability is in determining where, within a given city or community, specific environmentally impacting activities will take place. When various kinds of facilities, public and private, are sited, the decisions about where to site these facilities represent decisions concerning who will bear their costs or burdens. When a city decides to locate a solid waste transfer station, or a trash incinerator, or any of dozens of other kinds of facilities, the people who live in close proximity to these facilities often bear disproportionate burdens so that others can obtain the benefits. Similarly, when a company decides to locate an environmentally impacting industrial facility, the people who live closest to the facility bear disproportionate burdens so that others, including facility employees, can enjoy the benefits. In general, the burdens tend to be concentrated on those who live closest to the facility, and the benefits accrue to others or are distributed across a large geographically dispersed population.

Ostensibly, one of the impediments to achieving greater sustainability, as discussed in chapter 5, is the existence of the NIMBY syndrome. In general, the existence of the NIMBY syndrome is not necessarily favorable or unfavorable toward sustainability; overall, it is neutral with respect to achieving sustainability. Sometimes it makes siting environmentally friendly facilities difficult or impossible; often it makes siting environmentally threatening facilities difficult. When local opposition blocks a facility that promises to impose severe environmental degradation, this would seem to be a good thing for sustainability. When local opposition blocks a facility that would, on balance, improve the environment, this would seem to undermine achievement of sustainability.

The reason why facility-siting decisions, and their often associated NIMBY opposition, become issues of environmental justice has to do with the geographic concentration of minority or poverty status people within cities. Virtually any siting decision will affect one neighborhood more than another, and when the people who live in that neighborhood are largely African American, Hispanic, poor, or of another minority status, this raises issues of environmental justice, particularly when that neighborhood is already burdened by being in close proximity to other environmentally impacting facilities. The problem perhaps becomes even more complicated because of the fact that land and property values in minority communities or neighborhoods tend to be lower, and therefore look more attractive to developers just by virtue of the economics of specific projects (Portney 1991b). So, when a developer seeks to keep development costs down by proposing to site a new facility on cheaper property, which increases the potential return on the initial invest-ment, that is disproportionately likely to produce racially or ethnically inequitable results.

One central challenge, from the perspective of sustainable cities, is to find ways of making siting decisions more equitable, and to channel NIMBY opposition in order to increase the likelihood that facilities and activities that are good for the environment can be pursued, while those that are detrimental to the environment cannot. In some cities, usually independent of any type of sustainability initiative, per se, this has pro-duced a variety of different kinds of "fair share" proposals and policies. Fair share policies require some sort of equitable sharing of burdens across a city. Although few of the cities that have active sustainability initiatives incorporate fair share principles directly into these initiatives, such cities frequently have made efforts to make the burdens from eco-nomic development more equitable.

A second way that environmental justice issues relate to sustainability has to do with policy and program decisions made by city government, particularly concerning the distribution of benefits of those programs. Whether based on geography or on other characteristics, cities do not always provide the same level of benefits to all of their respective resi-dents. So decisions concerning how to deliver sustainability programs may have significant implications for the distribution of the benefits. For example, if a city enacts a green building program that applies only to new construction, then all areas of a city where there is no new construction would not reap any direct benefits. If a city designates specific areas as green space, and that green space is disproportionately located in predominantly nonminority parts of the city, this represents inequity. So the question is, what do cities do to monitor their degree of social and environmental justice by virtue of the distribution of city services? Some answers to this question are examined shortly.

Are Social Justice and Sustainable Cities Incompatible in Practice?

Much of this discussion on the relationship between social and environmental justice, on one hand, and sustainable cities as practiced in the United States assumes that there is some connection. Indeed, that is an assumption that seems to pour out of many of the nonprofit organizations that pursue sustainability in their respective cities. Yet not everyone accepts as a fundamental element of sustainable cities that greater environmental and social justice is, in practice, a goal. A dominant argument that has emerged in the literature that is critical of local sustainability initiatives, particularly the growth management side of it, is that sustainability is motivated by an expressed desire for middle-class White residents of cities to systematically undermine efforts at racial, ethnic, or socioeconomic integration. So the argument goes, the real purpose of local growth controls is to make it impossible for disadvantaged people to migrate into middle-class areas of cities. Thus, the argument is that, in practice, pursuing environmental sustainability is incompatible with the pursuit of social and environmental justice.

The argument that growth management has evolved as an instrument to foster exclusion is perhaps best represented by Frieden's (1979) "environmental protection hustle" thesis. This thesis, examined in the context of local growth control policies of cities and towns in the San Francisco Bay Area during the 1970s, suggested that an antigrowth coalition emerged to protect the status quo under the guise of environmental protection. In his words,

This coalition against homebuilding consisted of suburbanites who feared it would bring higher taxes and damaging social consequences, environmentalists concerned about the impact of growth on the natural landscape, and local government officials sympathetic to these views. (Frieden 1979, 3)

Frieden argues that such coalitions were able to drastically reduce the number of new homes that were built, thus denying homeownership to thousands of people and families who were thought to be "undesirable." To Frieden, local growth control policies are nothing more than the modern equivalent of now illegal exclusionary zoning practices of the 1950s and 1960s, and environmental rationales for these policies is nothing more than political cover to hide the true motivation of their advocates. To him, the reduction in the number of housing units built as a result of environmental concerns represents an unjust and inequitable result. Hence, the achievement of sustainability, at least in part through local growth limits, conflicts with efforts to become more socially and environmentally just.

This view of growth controls as the vehicle by which middle-class White communities are kept White and middle-class stands in sharp contrast to the apparent intent of achieving greater social and environmental justice as manifest in many sustainable cities initiatives. To Frieden and others, growth limits are neither necessary to achieve, nor compatible with, social and environmental equity. Yet, as noted earlier, to many advocates of urban sustainability, the achievement of greater equity is an integral part of sustainability itself. Other analyses raise questions about whether the assumptions underlying the "environmental protection hustle" thesis are accurate (Kee and Molotch 2000; Lewis and Neiman 2001; Neiman and Loveridge 1981). Which view is a more accurate description of the relationship between sustainability cities initiatives and equity may depend on how these goals are pursued in specific cities. Moreover, as discussed in chapter 4, sustainability initiatives in cities rarely advocate or endorse growth limits per se, opting instead to pursue growth management. In other words, they try to encourage certain kinds of growth rather than trying to stop it altogether. Presumably, the extent to which growth occurs in sustainable cities allows opportunities to avoid creating new inequities.

Social and Environmental Justice Issues in Sustainable Cities

How do issues of social justice and environmental equity manifest themselves, if at all, in cities' sustainability initiatives? Which cities have elected to address these kinds of issues, and which have steered clear of them? There are certainly many more of the former than the latter. Yet a brief look at cities' experiences provides some sense of how much variation there is. Moreover, it provides a picture of how cities can address issues of equity without articulating them as such. Frequently, efforts to address equity issues are embedded in general quality of life concerns or, as illustrated in the works of Kline (1995a, 1995b) in economic security issues.

Almost without exception, to the extent that equity is part of cities' sustainability initiatives, it is incorporated into their sustainability indicators. To the extent that any city articulates social or environmental equity concerns at all, that articulation is invariably contained in that city's sustainability indicators, whether as part of an indicators project, per se, or as part of cities' strategic or comprehensive plans. Sometimes indicators that are related to equity issues attempt to directly measure the distribution of some community characteristic, such as the degree of income inequality or the unemployment rate for people of different races; more commonly, such indicators focus on characteristics of specific population groups, such as the number of homeless people, or the services that would be targeted to those groups, such as the percentage of housing that is set aside for low-income people. Very often they are contained in concerns about specific "vulnerable" populations of people, such as the elderly or children.

Perhaps most prominent among cities that have developed such indicators are a number on the west coast, including Seattle and San Francisco. Other such cities that have equity-related indicators, away from the west coast, include Jacksonville, Florida, and Austin, Texas. As is clear, the kinds of indicators that have been adopted to monitor equity issues are much more varied across cities than those in many other areas, and indeed, one might question whether they are adequate to capture the broad concept of environmental justice. And as with cities' efforts to build community, even cities with equity indicators rarely provide any prescriptions for what the city can or should do to make things better.

Sustainable Seattle made explicit efforts to incorporate equity issues into its indicators project. According to a review of indicators projects provided by Global Cities Online (2000), the vast majority of social justice and equity indicators developed in any city were developed by

Sustainable Seattle as part of its indicators project. As of 1995, among the forty or so indicators eventually settled on in this initiative, there are a number that attempted to tap some element of social or economic equity or justice. For example, as an indicator of local economic performance, a measure of the distribution of personal income was developed, where growing income inequality was interpreted as a trend away from becoming more sustainable; and an indicator of the proportion of children living in poverty, where a higher proportion would represent a trend away from greater sustainability. Another equity indicator in Seattle focused on the ethnic diversity of public school teachers, where greater diversity is taken as consistent with sustainability. And yet another indicator focused on the local juvenile justice process, a measure of the disparate processing of children of different ethnic or racial backgrounds in the juvenile courts (AtKisson 1999, 356–357). Perhaps even more important than the equity indicators in Seattle is the incorporation of equity issues into the city's Comprehensive Plan. As part of its Human Development Element, this plan strategically outlines policies designed to address poverty, equity in health care, safety, and related issues (City of Seattle, 2002).

San Francisco's sustainability initiative contains numerous elements related to equity and social justice. Most of the effort there focuses on environmental justice, although the city's five-year Sustainability Plan uses language that interchanges social justice and environmental equity. San Francisco outlines five specific environmental justice goals: (1) to establish meaningful participation in the decision-making processes that affect historically disadvantaged communities of the City; (2) to create a vibrant community-based economy with jobs and career opportunities that allow all people economic self-determination and environmental health; (3) to eliminate disproportionate environmental burdens and pollution imposed on historically disadvantaged communities and communities of color; (4) to create a community with the capacity and resources for self-representation and indigenous leadership; and (5) to ensure that social and economic justice are established as an integral aspect of environmental well-being and sustainability (City of San Francisco 1996). These goals were accompanied by three specific indicators of social and environmental justice: the mean income level of people in historically disadvantaged communities (an increase is taken as a sign of movement toward greater sustainability); the proportion of environmental pollution sources in historically disadvantaged communities compared to other San Francisco communities (a decrease reflects greater sustainability); and the level of participation of historically disadvantaged communities as a whole and their representatives in city decision-making processes (an increase represents greater sustainability). Because there are no readily available sources of data for two of the three indicators, and for other measures of equity, San Francisco has not yet been able to track progress in achieving greater equity.

Each of these goals is linked to specific sets of programmatic prescriptions for the City. For example, the third goal, to eliminate disproportionate environmental burdens and pollution, was addressed through an effort to identify which communities have been disproportionately burdened, and to scrutinize any proposed projects in those communities to ensure that no new environmental burdens are created. Although Sustainable City San Francisco, a nonprofit organization, was successful in getting the city to officially adopt the plan it developed, since its adoption in 1997 it is difficult to find where in the operation of the city's governmental agencies these goals and programs are being vigorously pursued.

Jacksonville has been engaged in developing its Quality of Life indicators project for a relatively long period of time, starting with its initial effort in 1985. Developed as a joint effort between Jacksonville Community Council, Inc., and the Jacksonville Chamber of Commerce, the Quality of Life indicators represent a collection of dozens of measures organized in nine different substantive areas (Jacksonville Community Council, Inc., 1999). One of these areas is referred to as "Social Environment," and within this category of indicators, at least two deal directly with issues of equity. The indicators project never uses the terms "equity" or "social justice," but the social environment indicators seems to represent an attempt to tap elements of these concepts. Among the two indicators that would seem to reflect equity concerns most directly are one attitudinal indicator, measuring the proportion of the city's population that believes racism to be a problem, and one objective indicator, measuring the number of employment discrimination complaints filed with the Jacksonville Equal Opportunity Commission. Reductions in both of these measures would be consistent with becoming more sustainable. After experiencing steady increases in the former from 1985 to 1992, as measured in an annual public opinion survey, there have been slow but steady declines since 1993. Except in 1998, there have been substantial declines in the number of employment discrimination complaints since 1983.

Although only these two indicators would seem to represent direct measures of equity, there are four additional indicators that indirectly bear on equity. These indicators include measures of the size of particular at-risk populations, such as the number of substance-exposed newborns per 1,000 live births, the number of substantiated reports of child abuse and neglect per 1,000 children under the age of 18, and the number of resident live births to women under 18 years old per 1,000 live births. Presumably, sustainability would call for reductions in each of these over time, and this was the observed experience for all three indicators until fairly recently.

The case of the greater Austin area provides an example of an effort that attempts to fairly directly measure some kinds of equity. Although the indicators project there is a multicounty effort, it does illustrate the use of indicators of equity. In late 1999, the Sustainable Indicators Project of Hays, Travis, and Williamson Counties, the voluntary organization covering the greater Austin area, adopted four central indicators of equity in: education; law enforcement; access to investment capital; and achievement of leadership positions (Sustainability Indicators Project of Hays, Travis, and Williamson Counties 2000). The stated goal is to provide "equal access to justice, education, and economic advancement without regard for race or ethnicity." Equity in education is measured by the proportion of the three counties' students attending "exemplary schools," by the race/ethnicity of the students. Presumably, any difference in the proportion of students in each of a number of race/ethnicity groupings represents inequity. The 2000 report suggests that the gap between Black and White students has grown substantially since 1994–95. In that year, 1.1% of African American students and 12.2% of White students attended exemplary schools, for a difference of 11.1%. By 1998–99, 5.3% of African American students and 31.7% of White students attended such schools, producing a gap of some 26.4%. The gap for Hispanic students went from 10.9% to 19.8% over the same period (Sustainability Indicators Project of Hays, Travis, and Williamson Counties 2000, 13).

Equity in law enforcement is measured in the Austin initiative by the ratio of the percentages of youths' encounters with law enforcement (arrests) to youths' percentages in the overall population for each race/ethnicity group. A ratio over 1.0 means that this particular group has more encounters with law enforcement than their numbers in the general population. Overall, this indicator suggests very little change over the period from 1994 to 1998. In 1994, Black youths had a ratio of 1.97 (Black youths were almost twice as likely to be arrested as their proportion in the population), and in 1998, the ratio was 1.79, suggesting a slight improvement. For Hispanic youths, the ratio went from 1.27 to 1.15 over the same period, again suggesting slight improvements (Sustainability Indicators Project of Hays, Travis, and Williamson Counties 2000, 14).

Equity in access to capital is measured by the proportion of individual home loan applications that were approved by race/ethnicity group. For this indicator, only two data points are reported, for 1997 and 1998. Although no effort is made to report the success rates for home loans controlling for the income of the applicant, in both years African American and Hispanic applicants were only about half as likely to have their applications approved as White applicants. In 1998, nearly 70% of White applications were approved, while only 37% of African American and 37% of Hispanic applications were approved (Sustainability Indicators Project of Hays, Travis, and Williamson Counties 2000, 15).

Equity in leadership positions is measured by the ratio of the percentage of persons who are ethnic minorities or women in regional leadership positions (public officials or chief executives) to the percentages of these groups in the population. A ratio of 1.0 would mean that a particular group is represented in leadership positions in the same proportion as its numbers in the population. Here, information is reported only for the year 2000 because no systematic effort had been made to collect if before the development of the indicators project. The results showed that Whites had a ratio of 1.31 (White people were more likely to be in leadership positions than their numbers in the population), while Blacks had a ratio of .74 and Hispanics of .51, suggesting substantial inequity (Sustainability Indicators Project of Hays, Travis, and Williamson Counties 2000, 16).

The three-county indicators project makes a relatively impressive effort to incorporate issues of equity into its sustainability initiative. Few indicators projects do more. Perhaps equally telling, however, is that the sustainability initiative operated by the City of Austin makes a much less explicit effort to incorporate equity issues. In the City effort, much of the sustainability effort is focused on the environment and "smart growth" initiatives, although there are specific projects that do take equity into consideration. For example, in assessing the relative merit of, and making funding decisions on, various capital improvement projects for the City, social justice (the degree of equity and diversity provided by each project) represents one of several considerations in a "multi-attribute decision matrix" (City of Austin 2000). One might argue that this represents a more impressive effort than found in many cities simply by virtue of the fact that this is a specific program designed to advance the cause of equity. It goes beyond the simple definition of indicators, trying instead to achieve greater equity. Whether it has accomplished this goal is not clear.

Social and Environmental Justice: A Summary

While many advocates of sustainable cities suggest that issues of social and environmental justice are integral to the definition of sustainability, whether cities do in fact incorporate such issues into their respective sustainability initiatives is highly variable. Conceptually, it is possible to imagine a sustainability effort that does not include special attention to issues of equity and justice, and many cities seem to implicitly adhere to such a conception. Moreover, there is, at least conceptually, a clear tension between pursuing justice elements of sustainability and adopting local growth controls, where it is at least possible that the latter impede the former. Empirical evidence on this is, at best, sketchy.

Despite the fact that a number of cities have elected to address issues of social justice and environmental equity in their sustainability initiatives, most cities have not done so. The reasons why some cities address these issues and some do not can only be treated as a matter of speculation here. Certainly, cities that are ideologically predisposed to aggressively pursue local programs and policies that tend to benefit racial and ethnic minorities in general will use their sustainability efforts to advance this cause. And cities that are not so predisposed will not. Yet some cities that would seem likely, at least from an ideological perspective, to address issues of equity, such as Santa Monica, do not. And others that would seem unlikely to address these issues, such as Austin, elect to do so. It probably does not hurt the cause of equity that Austin is a college town and the state capitol, yet these characteristics, by themselves, would not seem adequate as explanations.

Even those cities that have elected to incorporate equity considerations into their sustainability initiatives have done so in only a superficial way. The indicators developed in these cities are but the most basic and perhaps limited measures of how much equity exists. Simple measures of income inequality, of differential health of at-risk populations, and of law enforcement outcomes may represent a good start, but do not capture the essence of environmental or social justice. None of the cities examined here made an effort to develop indicators or measures of differential exposures to environmental contamination. None of the cities attempted to monitor the differential environmental burdens of some groups of people over others. Only a few cities incorporated measures of the distribution of city services, such as San Francisco's efforts to equalize access to recycling services across all neighborhoods in the city. And none of the cities has created specific programmatic initiatives to improve the indicators of equity. It may be unreasonable for any city to initiate an effective program for affecting the level of income inequality, but programmatic efforts to deal with equity issues are generally not part of local conceptions of sustainability. Suffice it to say that if equity issues are important conceptual components of sustainability, then sustainable cities initiatives in the U.S. do not seem to take it very seriously. Perhaps it would be more accurate to say that as practiced in most cities, equity issues do not appear to be integral parts of cities' definitions of sustainability.

7

Cities That Take Sustainability Seriously?
Profiles of Eight Cities

There is no city in the world that does everything that it could conceivably do to become truly sustainable. Stated perhaps more accurately, no city has yet fully come to grips with the full range of activities and issues that it could confront or would have to confront to begin moving toward becoming more sustainable, even given limitations on knowing what becoming more sustainable means. Nevertheless, given the preceding analysis, it is possible to look at the activities and initiatives of cities, and to begin making some preliminary judgments about the degree to which these activities, taken as a whole, might seem to constitute serious efforts at becoming more sustainable. At a minimum, a look at current "best practices" in a variety of cities will delineate the range of possibilities cities have been able to achieve given their own respective political, economic, and social constraints. Of course, any effort to make such judgments represents a slippery slope since, as discussed in chapter 2, there are no clear-cut or universally agreed upon standards and criteria to use to measure this.

The preceding analysis carried implications for the kinds of elements that must be present or that would appear to be important in any serious sustainability initiative. The purpose of this chapter is to examine the broad range of activities in specific cities, taking a holistic rather than atomistic view. In short, are there specific cities that have addressed the myriad issues that presumably need to be addressed? Are there cities that seem to be on the right track given what we think we know about what the right track is? Are there cities that seem to be "getting it right?" In the context of the day-to-day life in these cities, what does it mean to "get it right?" To the extent that specific cities do seem to be getting it right, what are some of the processes and difficulties they have had to confront and perhaps overcome? Are there common themes and traits that characterize these cities? As difficult as these questions are to answer, this chapter makes a concerted effort to examine them. The title of this chapter, however, is in the form of a question because, even in the final analysis, it is difficult to know with any degree of certainty whether these cities have indeed gotten it right.

The analysis that follows focuses on eight cities: Austin, Boulder, Chattanooga, Jacksonville, Portland, Santa Monica, San Francisco, and Seattle. These may not be the only cities that seem to be getting sustainable cities right, and indeed there are a number of other cities that could be profiled here as well. Cities such as San Jose, Tampa, and Scottsdale, to name a few identified in chapter 2 and described more fully in chapter 8, also rank highly and certainly could have been included here. But these eight cities do represent a selection of cities that have done many innovative things to take sustainability seriously, at least within some broad range. This is not to say that these eight cities are all the same. Indeed, the discussions that follow make it clear how different they are in terms of their commitment to, and seriousness about, sustainability. One could even make the argument that some of these cities compare so unfavorably to others that they cannot be said to take sustainability very seriously at all. As strong as their programs are, Austin, Jacksonville, and even San Francisco, for example, have not been able to achieve quite the level of programmatic success as that found in some of the other cities. Yet all eight of the cities profiled here are engaged in the pursuit of sustainability at a far higher level than the average U.S. city, where little thought or action has been given to such issues.

Additionally, there is no way to know what the underlying universe of cities that take sustainability seriously is. There is no existing database containing information about the whole range of sustainability activities in which cities may be engaged. There is no existing database containing information about what all major U.S. cities are doing, if anything, in practice. So there is no way that an argument can be made that these cities are in any way "representative" of some larger universe of cities. Moreover, these cities were selected, in fact, because they appear to be very different from many, if not most, other cities when it comes to sustainability. Indeed, one of the purposes of this chapter is to make the case that these cities are somewhat unique in that respect. Suffice it to say that the purpose of this analysis is to provide some very preliminary answers to the questions posed with the understanding that the answers provided here are at least partly dependent on the specific cities selected for discussion.

Eight Cities That Seem to Take Sustainability Seriously

The eight selected cities profiled here provide a glimpse into the actual workings of cities' sustainability initiatives. Although their programs differ, they share the fact that sustainability plays a prominent role in the cities' activities. Yet this may be one of the few ways in which the cities are similar. A simple look at table 7.1 shows that the eight cities capture a range of demographic and population characteristics. Five of the cities, Austin, Jacksonville, San Francisco, Seattle, and Portland, are fairly large, with populations of over half a million. Two of the cities, Santa Monica and Boulder, are fairly small, with populations of just under 100,000. And Chattanooga is also on the small side, with a population of about 150,000. Two of the cities, Chattanooga and Santa Monica, experienced population loss between 1980 and 1990, the decade just prior to their foray into the realm of sustainability, and four cities, Austin, Jacksonville, Portland, and Boulder, experienced substantial population growth over the same period. Across all eight cities, the average 20-year growth rate of 28.7% may seem substantial, but of course represents less than 1.5% annual average growth rate even without compounding.

Only Chattanooga has a fairly large African American population (36.1% as of 2000), and only Austin and Santa Monica have substantial Hispanic populations (28.5% and 13.4%, respectively, as of 2000). The cities also vary widely in terms of their geographic footprints, with land areas ranging from 840 square miles in Jacksonville to just over 8 square miles in Santa Monica. This translates into gross population densities (number of people per square mile) of a low of 920 in Jacksonville to a high of over 17,000 in San Francisco. The overall average population density for these eight cities is a little under 6,000 people per square mile. Two of the cities, Boulder and Santa Monica, have relatively high median family incomes, and two are relatively low (Chattanooga and

Table 7.1
Population characteristics of eight profiled sustainable cities

City	Population 2000	Population 1990	Population 1980	Population change, 1980–1990 percent	Population change 1980–2000 percent	Land area (square miles)	Population density 2000
Santa Monica	90,777	86,905	88,314	−1.6%	+2.8%	8.3	10,937
Boulder	96,727	95,395	83,908	+13.7%	+15.3%	25.4	3,808
Chattanooga	147,790	152,488	169,565	−10.1%	−12.8%	124.0	1,192
Portland	508,500	437,319	366,383	+19.4%	+38.8%	144.8	3,512
Seattle	539,600	516,300	493,846	+4.5%	+9.3%	84.0	6,424
Austin	642,994	465,622	345,496	+6.6%	+86.1%	228.6	2,813
Jacksonville	772,544	672,971	571,003	+17.9%	+35.3%	840.0	920
San Francisco	801,377	723,959	678,974	+34.8%	+18.0%	46.7	17,160
8-city average	450,039	393,870	349,686	+17.0%	+28.7%	187.7	5,846

City	Percent Black 2000	Percent Hispanic 2000	Median family income 1990	Percent families in poverty 1990	Percent employed in manufacturing 1990	Average unemployment rate 1994–1998	Per capita local government expenditures 1990
Santa Monica	3.8%	13.4%	$51,085	5.7%	9.5%	8.0%	$1,685
Boulder	1.2%	8.2%	$46,208	7.4%	11.1%	3.6%	$905
Chattanooga	36.1%	2.1%	$27,487	14.4%	18.0%	3.5%	$1,600
Portland	6.6%	6.8%	$32,424	9.7%	15.1%	4.3%	$909
Seattle	8.4%	5.3%	$39,860	7.4%	13.3%	4.4%	$1,182
Austin	11.5%	28.5%	$33,481	11.5%	11.2%	3.1%	$1,167
Jacksonville	29.0%	4.2%	$33,303	9.9%	9.4%	5.2%	$1,192
San Francisco	7.8%	14.1%	$40,561	9.7%	9.2%	4.4%	$2,887
8-city average	13.1%	10.3%	$38,051	9.5%	12.1%	4.6%	$1,441

Portland). Chattanooga and Austin have relatively high poverty rates, with Santa Monica's less than half that of these cities. None of the cities appear to be greatly dependent on manufacturing industries as a base of employment, with Chattanooga and Portland being the most dependent. Average unemployment rates ranged from a little over 3% in Austin to a high of 8% in Santa Monica. And government expenditures varied considerably, ranging from a little over $900 per person in Boulder and Portland to nearly $3,000 in San Francisco. In short, none of the basic demographic and other city characteristics would seem to readily differentiate these eight cities.

Austin   In conjunction with Travis county, where Austin is located, and the surrounding counties, Williamson and Hays, which constitute the greater Austin metropolitan area, the city of Austin has created an important sustainability initiative. It is important not just for its programmatic and policy elements, but also because of its accomplishments in a state that does not make the pursuit of sustainability easy. Austin, the state capitol of Texas, is a fairly large and rapidly growing city with a current population of nearly 643,000 people in a metropolitan area of over 1.1 million. The city has nearly doubled in population since 1980, when its population was about 346,000. Much of its growth is attributable to a combination of the presence of the University of Texas's main campus, the growth of a variety of technology-based companies, and an aggressive annexation effort. The major employers other than the University and government are Dell Computer, Motorola, and IBM. In geographic size, Austin is the second largest of the eight cities profiled here, with land area of over 228 square miles. This gives the city a gross population density of a little over 2,800 people per square mile, well below the average for the eight cities. The city has a "weak mayor" form of government, and is governed by a seven-member city council, all elected at large. The council selects the mayor from among its ranks. The city's chief executive is a full-time city manager who is appointed by and serves at the pleasure of the council.

The sustainability initiative really has two related parts. The first part is associated with the "Sustainability Indicators Project of Bastrop, Caldwell, Hays, Travis, and Williamson Counties" (Central Texas Indicators 2001 2001), the counties that constitute the metropolitan area for

Austin. This indicators project, first developed in 1997, focuses on a comprehensive array of some 42 indicators in four major categories—community and children (including public safety, education and education equity, civic engagement, and volunteerism), workforce and economy (including the cost of government, unemployment, the cost of living, household income, availability of affordable housing, and diversification of the employment base), health and environment (including air quality, hazardous materials, energy use, and solid waste), and land and infrastructure (including public open space, density of new development, and time spent commuting). Figure 7.1 provides a reproduction of a page from the 2001 indicators report showing the content and form of the effort to measure progress toward improving the amount of hazardous materials released into the environment. The indicator presented in this figure, from the Toxics Release Inventory, shows a reduction of hazardous materials between 1992 and 1997, and an increase since that time. It also provides a comparison of the three counties and with other counties in Texas and other states.

Additionally, the City of Austin operates its own indicators project. Called "Sustainable Community," this project has since 1999 become part of a city-wide initiative to manage city government by results, and to provide comprehensive reporting on government performance. The city's sustainability indicators cover public safety (fire and medical services, police, and the courts), youth, family and neighborhood vitality (including health services, housing, libraries, and parks and recreation), sustainability (traffic and road maintenance, air quality, recycling and waste diversion, drinking water quality, lake and stream quality and water conservation, energy conservation, and inspections and site plan/subdivision review), and affordability. These indicators are used as measures of the performance of local government, and results are reported annually in the City of Austin "Community Scorecard." Much of the contents of the Scorecard and indicators themselves are based on the semi-annual "Voice of the Customer Survey," a telephone-based survey of about 500 randomly selected city residents. The sustainability initiative in Austin goes well beyond the development and use of these indicators. A brief review of some of the programs and projects developed in Austin provide a sense of how extensive Austin's sustainability initiative is.

Hazardous Materials

30

Current State
The overall release and transport of hazardous materials in Central Texas is not decreasing significantly.

Ideal State
Central Texans are not exposed to harmful levels of toxic or hazardous materials.

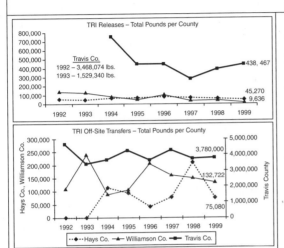

Measure
The Toxics Release Inventory.

Data Source
Data are collected by the U.S. Environmental Protection Agency. The Toxics Release Inventory (TRI) is a federally mandated accounting of specified chemicals released or transported from specified facilities. Environmental releases are point-source discharges such as air emissions, surface water discharge, underground injection, land application (which includes landfill), and water-based treatment. Off-site transfers are the movement of specific hazardous materials from one site to another for additional treatment, storage, disposal, recycling, or burning (for energy recovery).

Findings
From 1992 to 1999, the three-county region experienced a significant overall reduction in the amount of TRI-listed toxic materials released into the environment. Travis County has consistently accounted for approximately 95% of the total transfers in the region, although Travis County's share of the total regional releases dropped from 95% in 1992 to 89% in 1999.

Context
The pounds of environmental releases in 1998 for other counties were as follows: Bexar County - 892,294; Tarrant County - 1,476,923; and Wake County, North Carolina (Raleigh) - 957,550. The pounds of off-site transfers in 1997 for other counties were: Bexar County - 1,415,144; Tarrant County - 2,807,287; and Wake County - 686,927.

Figure 7.1
Hazardous materials indicators from the Sustainability Indicators Project of Hays, Travis, and Williamson counties (Texas)

Conceptions of sustainability, particularly with respect to the environment, have filtered into many of the programs and policies pursued by the City. For example, as part of its smart growth efforts, and its land use planning, Austin uses a system to assess and rank proposed capital improvement projects according to their relative contributions to sustainability, as determined by a multi-attribute utility matrix (City of Austin 1999). The effect of assessments using this matrix is that it privileges proposed projects for city funding that rank higher in their contributions to sustainability. Moreover, private contractors and developers can, at their discretion, participate in a program whereby they apply the same utility matrix and submit an application documenting the ways in which their projects contribute to the sustainability goals of the city. After formal review and consideration by the city council, a particular development project might qualify for a substantial city "incentive package," which can include fee waivers, reimbursement of fees already paid by the applicant, and property tax abatements.

Another major component of Austin's sustainability effort is associated with its publicly owned electric utility, known as Austin Energy, and the recommendations of The Sustainable Energy Task Force. The sustainability efforts on energy use stem from the city council's resolution that 5% of the city's electricity should come from renewable sources by the year 2005. The utility operates a program that allows customers to elect the "GreenChoice" option where they receive electricity generated from renewable sources (wind, solar, and biogas from closed landfills) at a fixed rate guaranteed for ten years, while traditional customer's rates are not guaranteed and fluctuate with the market price of fossil fuels (Austin Energy 2001). The utility plans to construct new wind turbine generating facilities in West Texas to provide additional renewable sources to power up to 20,000 homes.

Additionally, the city has developed a "green building program" to provide technical information and guidance to developers concerning how to build more environmentally and energy-efficient construction. The green building program, which has residential, municipal, and commercial components, includes an effort to encourage developers to engage in smart development. The Austin conception of green building, as noted in the program description, is that it ". . . is based on a market-pull mechanism whereby the Green Building Program promotes green building practices, rates buildings that feature these practices, thus creating more demand from the public because these buildings are perceived as more attractive products for people to buy" (City of Austin 2001a). As of early 2001, the program has rated 1,800 homes, 1,400 apartment units, and 10 commercial buildings and has consulted on 85 other commercial projects.

Austin also engages in "transit-oriented development" land use planning to minimize reliance on private transportation (City of Austin 2000), has engaged in the development of a "sustainable purchases protocol" for the municipal government that sets standards for city purchases of good and services that are environmentally friendly (City of Austin 2001b) and a public private partnership, called Greater Austin@work, designed to foster economic development and job growth in sectors that produce fewer environmental impacts.

Chattanooga One of the most frequently cited examples of a city that has made enormous strides in working toward becoming a more sustainable place is Chattanooga. Located in Southeast Tennessee just above the Georgia border, it is a city of about 124 square miles with a current population of just about 150,000 in a metropolitan area of nearly 500,000 people. The city has a population density of about 1,200 residents per square mile. The size of Chattanooga's population has declined more than 10% over the past twenty years, after reaching a peak of nearly 170,000 in 1980. Chattanooga has a full-time mayor elected at-large, and a city council of nine members elected by districts on nonpartisan ballots.

Although Chattanooga has engaged in an impressive array of sustainability activities, there is little doubt that it "takes sustainability seriously" in a very different way from many other cities. Clearly, this city does not see sustainability as falling in the central purview of city government, nor does it see achievement of sustainable outcomes as being the direct result of local public policy and programs alone. Moreover, in Chattanooga there is an explicit linkage between what is good for business and economic development and what makes the city livable. Stated another way, Chattanooga's interest in sustainability is driven by a healthy dose of the pursuit of sustainable economic development. While livability issues are a central part of the motivation for pursuing sustainability, livability itself is an economic development issue. To many supporters of sustainability in Chattanooga, the way to achieve greater economic development is by making the City more livable, and pursuing sustainability is one way of improving the city's livability. This is seen in the central role played by the Chamber of Commerce, now an area-wide organization that has consolidated a number of programs within the city's region.

Chattanooga's sustainability efforts themselves evolved over a fairly long period of time, with a variety of foundations having been established starting in the early 1970s. The sustainability efforts as they appear today began in earnest with the "Vision 2000" initiative that grew out of the work of the Lyndhurst Foundation and the "Chattanooga Venture" initiatives it supported in the early 1980s. Vision 2000 set in motion an effort to involve a wide array of community members and leaders in a process to consider the problems and solutions faced by Chattanooga, and to define specific goals as ways of redressing these problems. This "visioning" process yielded six broad categories of goals, focusing on "people, places, play, work, government, and future alternatives." In the early 1990s, apparently as a result of perceived successes of the Vision 2000 initiative, the "ReVision 2000" initiative was undertaken, and it is this process that invigorated a focus on sustainability (Parr 1998).

Chattanooga began this process as a troubled city. With the unfettered growth of the steel, coal, and other industries during the 1960s, Chattanooga developed some of the most serious pollution problems of any city in the nation. Air pollution during the 1960s was the worst problem, with residents frequently required to drive their cars with their headlights on in the middle of the day because pollution had significantly reduced visibility (Graham 1999, 4). As L. Joe Ferguson, whose Chattanooga-based company Advanced Vehicle Systems, Inc. was created to build the City's electric bus fleet, noted about the state of air pollution in 1969, "Walter Cronkite announced on the news that we had the poorest air quality in the country, and boy, let me tell you we did" (Graham 1999, 5).

The Vision 2000 process, started in 1984, was born of deep concern over the economic decline of the city during the previous decade.

Although the deindustrialization of the Chattanooga area, paralleling the decline of the domestic steel industry, had brought significant improvements in air pollution, it also ushered in an era of economic decline and associated disinvestments in the city's economic base. In response to this decline, a number of organizations contributed to engaging the people of the city in a process to define new paths toward economic development. The issue became how to promote economic development without undermining the quality of life in the city. The consensus even among business leaders seemed to be that the city needed a healthy infusion of economic development, but not at the expense of the quality of life. No one wanted to return to the days when local industrial pollution undermined the quality of life. Out of this and related processes grew a series of specific projects and activities that, taken together, define the nature of sustainability in Chattanooga.

Far more than any of the other four cities profiled here, and more than most other cities commonly associated with sustainability, Chattanooga's initiatives are driven by economic development goals. Moreover, the city government, per se, is much less involved in the pursuit of sustainability than most of the other cities. Virtually none of the activities associated with sustainability in Chattanooga have been directed, administered, or spearheaded by a city agency of any sort—not by multiple city agencies working independently, and certainly not by any single agency or nonprofit organization. Programmatically, perhaps the single most important organization in the city is the Chamber of Commerce. As described later, the Chamber's role evolved to the point where it plays more of a central coordinator for sustainable development than any other organization or agency.

The county regional planning agency has played a larger role than the city government in that it has adopted more of a sustainability mind-set, but that role is still relatively minor because it has much less actual programmatic and budgetary authority than city agencies. Instead, the initiatives are the result of the independent or sometimes loosely connected activities of many different organizations, including a number of nonprofit corporations and foundations, and the Chamber. There is an "indicators of community well-being" project for Hamilton County spearheaded by the Metropolitan Council for Community Services, and nonprofit United Way organization (Metro Council 1998). Additionally, the ReVision 2000 process produced a series of goals that resemble, in some ways, the products of other cities' indicators projects. In short, if the litmus test were a well-developed strategic sustainability plan with fully delineated indicators, benchmarks, goals, and timetables, centrally-managed by a single administrative city agency, Chattanooga's initiative would not appear very serious when compared to many other cities, as described later. However, there are many features of the economic development and related processes in Chattanooga that do reflect a serious effort.

The sustainability initiative in Chattanooga today consists of a number of disparate pieces that when taken as a whole represent an impressive array of activities. Beginning with the Vision 2000 and ReVision 2000 processes described previously, the heart of many of the sustainability activities in Chattanooga is manifest in the nonprofit sector. The Chattanooga Greenways program has worked to create a 75-mile network of protected corridors of open space linking many parks, recreational areas, and attractions within the metropolitan area. Started through the efforts of the Trust for Public Land, a national nonprofit organization, since 1994 the Greenways program has engaged a number of local and neighborhood associations and city agencies in a partnership to acquire new lands and to protect existing areas that become designated as part of the Greenway. Currently, there are eight pieces to this greenway—Lookout Creek, Lookout Mountain, North Chickamauga Creek, Tennessee Riverpark, South Chickamauga Creek, The Blue Blazes Historic Trail, the Brainerd Levee, and South Chattanooga.

Another important part of Chattanooga's initiative is Chattanooga Neighborhood Enterprise, Inc. (CNE). This nonprofit corporation, founded in 1986, concentrates on providing affordable and low-cost housing. Like affordable housing initiatives in other cities, CNE provides mortgage and rehabilitation finance services, a small amount of development, and homeowner education programs. Again, economic development is a central concern of this enterprise. One description of the program notes "CNE brings money into the Chattanooga area and builds equity in people and property. It turns many users of public services into contributors through payment of property taxes" (Chattanooga Area Chamber of Commerce 2000). Clearly, this nonprofit is more focused than counterparts elsewhere on doing development projects, particularly in housing, but in this case specifically with an eye toward achieving greater equity (Pierce and Johnson 1998).

Another large piece in the Chattanooga puzzle has been RiverValley Partners, Inc., once described by the Chamber of Commerce as a public–private, nonprofit organization dedicated to implementing an economic development strategy that attracts new investment in the city while explicitly seeking to attain quality of life goals. RiverValley Partners was created in 1993 after a merger between two predecessor organizations, the RiverCity Company, a nonprofit group established in 1986, and the Partners for Economic Progress, which produced "Target '96," the City's first explicit environmental strategic plan. RiverValley Partners and its predecessors sponsored numerous economic development projects throughout the city and its county, including the Tennessee Aquarium, opened in 1992, the Riverwalk, and the Tennessee Riverpark. By 2001, RiverValley Partners was folded into a larger initiative of the Chamber of Commerce, and essentially became part of the Chattanooga Regional Growth Initiative oriented around what the Chamber calls "cluster development." Cluster development focuses on strategically trying to attract businesses in specific industries or sectors of the economy. For example, the Chamber has designated health and hospitals, confectionary and baked goods, medical devices and health services, textiles and floor coverings, and a number of other clusters as targets for development. One of these clusters is oriented around the electric vehicle initiative started for the purpose of providing zero-emission buses for the city's public transit system, but otherwise the clusters would appear to play no particular role in sustainability. Indeed, the criteria used to determine whether a cluster should be delineated all relate to economic competitiveness and comparative economic advantage, and sustainability criteria are certainly not prominent or obvious.

Perhaps the most comprehensive piece of the sustainability puzzle is the "Futurescape Community Planning Process" initiated by the Chattanooga Hamilton County Regional Planning Agency (CHCRPA 1997). This regional planning agency, which has responsibility for state-mandated, county-wide planning, probably internalizes sustainability to a higher degree than can be found within the city of Chattanooga itself. This can be seen in the Futurescape planning process, and in its mission statement and organizational principles, outlined in table 7.2. When the Regional Agency states its mission as creating a "... comprehensive vision ... that enhances the quality of life by integrating growth with the conservation of resources," and articulates principles suggesting that "planning should reflect the integration of comprehensive economic, social and environmental factors," and "plans should promote the wise use of existing resources without destroying our options for the future," there is at least a core of the values common to virtually all definitions of sustainability.

The Futurescape planning process, which is now part of the state-mandated, county-wide plan called the "2020 Plan," consists of three major elements. It began in 1996, where several thousand residents from the ten municipalities in the County (of which Chattanooga is the largest) participated in that is called the "Visual Preference Survey." In this survey, residents were asked to respond to pictures of different types, locations, and characters of potential new development, and to articulate what kinds of development they preferred over others. The second element took the results of the visual preferences survey and built them into a comprehensive plan for development. The third element was to revise various regulations, especially zoning ordinances and building codes, in order to guide development in ways consistent with the first two elements.

Perhaps the most basic difference between the sustainability efforts in Chattanooga and the other cities profiled here relates to the role of certain quality of life goals, particularly those associated with empowerment, and environmental and social justice issues. In Santa Monica, Seattle, Boulder, and to some extent Portland, the sustainability initiatives incorporate *explicit* efforts to engage a broad spectrum of residents in the process, and to achieve higher levels of equity in allocating environmental risks. These kinds of efforts and concerns are largely absent from the initiatives in Chattanooga. There is certainly no official recognition of them as important elements in their sustainability pursuits. Clearly, citizens' organizations have played an enormous role in making many of the sustainability activities happen in Chattanooga. The Vision 2000 and ReVision 2000 processes were characterized by significant amounts of resident participation—perhaps as many as 1,700 residents participated in the former, and 2,600 in the latter (Parr 1998; Pierce and

Table 7.2
Statement of mission and organizational principles, Chattanooga Hamilton County Regional Planning Agency

Mission statement
Our mission is to provide a comprehensive vision and guide for the community that enhances the quality of life by integrating growth with the conservation of resources. This vision will include both short- and long-range goals and strategies that public and private community leaders can use to implement these objectives.

Organizational principles
It was recognized that good planning has certain qualities or principles of excellence. Any planning activity should be conducted with these principles in mind. The following principles of good planning will help guide the work of the Planning Commission.
1. Planning should be proactive and visionary.
2. Planning should reflect the integration of comprehensive economic, social, and environmental factors.
3. Plans should promote the wise use of existing resources without destroying our options for the future.
4. Planning should recognize the importance of the diversity in our community, including its people, cultures, values, places, and natural resources.
5. Citizen involvement in the planning process is essential.
6. A high ethical standard, free from conflicts of interest, should permeate the planning process.
7. Planning should seek to find a balance between what is good for the community as a whole, and the rights of citizens as individuals.
8. Plans should incorporate realistic implementation components that define specific areas of responsibility.
9. Planning should recognize the importance of the central city to the economic and cultural vitality of the entire community.

Source: Chattanooga Hamilton County Regional Planning Agency (2000). <http://www.chcrpa.org/statement_of_organizational_prin.htm>

Johnson 1998)—and many neighborhood associations were instrumental in advocating for specific projects. As a media relations specialist for the Chamber of Commerce put it, ". . . Chattanooga is pretty conservative place. The first reaction people have is: don't go to government leadership, it's a waste of time. There has been a void in public leadership that was filled by public participation . . . citizens organizations have definitely pushed government and made things happen that probably wouldn't otherwise have happened" (Chattanooga Area Chamber of Commerce 2000). Yet, as important as public participation has been, none of the sustainability initiatives has outlined specific goals and programs to promote participation or to monitor how much participation actually takes place.

Another important difference relates to the importance of nongovernmental organizations. Almost all of the sustainable cities profiled here have one or two major nonprofit organizations that have played a role, sometimes instrumental, in the sustainability initiative. Yet in Chattanooga, nonprofit organizations are far more prolific, and along with the Chamber of Commerce, play a much larger role vis-à-vis the city government itself, than is the case in other cities.

The critical issue for Chattanooga's sustainability initiative is whether, and to what extent, a process largely dominated by the business community, in this case through the Chamber of Commerce, can take sustainability seriously. Clearly the rhetoric suggests that it can. In a 1997 presentation at a Chautauqua Conference on Regional Governance, the President of the Chattanooga Chamber of Commerce suggested that the reason why his organization became interested in sustainability was "because the old strategy—low taxes, low cost of land and construction, low wages, and cheap power—wasn't working anymore. We believe some things must grow—jobs, productivity, income and wages, profits, capital, and savings, information, knowledge, education. And that others must not—pollution, waste, poverty, energy and material use per unit of output" (Pierce and Johnson 1998). According to his view, the role of the Chamber of Commerce is to help make possible a particular vision where, as he stated it, "we are going to build the future of Chattanooga by balancing the economy, ecology, and equity" (Parr 1998). Yet the question remains whether the actions and results in Chattanooga are anything more than strategic economic development by another name.

Is Chattanooga simply engaged in understanding and using its comparative economic advantage for the purpose of achieving economic development? Or is the Chamber facilitating the balancing of economic growth with environmental concerns of the sort that is promoted by smart growth advocates. By 2001, the Chamber's emphasis on cluster development with little or no mention of other missions certainly suggests that the goal is more growth than smart growth.

Seattle   The city that stands at the top of virtually every list of sustainable cities, particularly those cities that have developed major indicators projects, is Seattle. Located in Western Washington on the Puget Sound, Seattle is the largest city in the state, with a population of about 539,000 in a metropolitan area of over 3 million people. It is a city of approximately 84 square miles, giving it a population density of 6,453 people per square mile. Although the population of Seattle declined from about 530,000 in 1970 to a low of about 490,000 in 1980, it has experienced steady growth since then. Seattle has a mayor–council form of government, with the full-time mayor, who serves as the city's chief executive officer, elected at-large every four years. The City Council consists of nine members all elected at-large.

The sustainability efforts in Seattle are found in both the nonprofit and governmental sectors, but the City's initiative is primarily associated with the activities of a single organization. "Sustainable Seattle," the name of this initiative, was founded in 1990 and began its operations in 1991. The Sustainable Seattle initiative was conceived of as an operation defined as a nonprofit organization (it currently operates under a 501(c)3 tax status designation), and its origins are significantly more grassroots than are the sustainability initiatives in most other cities. The organization, which is governed by a board of directors and has a small staff, describes itself as a "volunteer-based civic network and forum . . . with a focus on the metropolitan city/county area" of Seattle. Its mission is to "protect and improve [the Seattle] area's long-term health and vitality by applying sustainability to the links between economic prosperity, environmental vitality, and social equity" (Sustainable Seattle 2000a). Out of the mission has grown some six specific goals that include influencing individual and collective local actions that are thought to move the city towards greater sustainability, preparing and publishing sustainable community indicators, providing extensive information about sustainability to residents and local leaders, putting issues of sustainability on the agendas in peoples' homes, neighborhoods, places of employment, schools, and in civic life generally, providing an open forum for "cross-community dialog" on issues of sustainability, and to serve as a general resource center.

Perhaps the key defining characteristic of Sustainable Seattle, the characteristic that gave this initiative national notoriety and attention, is its Sustainable Indicators Project (AtKisson 1996). The Indicators Project developed a wide array of indicators for the city, a sample of which is shown in box 2.2 in chapter 2. This project's notoriety grew not only out of the resulting indicators themselves, but also out of the processes that were used to produce them. As described in chapter 2, and consistent with Sustainable Seattle's goal of providing a cross-community forum for discussion of sustainability issues, the Indicators Project sought to derive its indicators through a fairly participatory process.

As impressive as the Sustainable Seattle organization and its Indicators Project have been, they tell only part of the sustainable city story in Seattle. The organization certainly articulates a goal of influencing local collective actions, but the organization does not itself have any sort of legal authority for adopting or implementing policies that promote sustainable growth. The organization can use (and has used) its sustainability indicators as a political weapon by, for example, reporting the nonattainment of specific environmental goals, but it cannot directly establish public programs that will ensure that the goals are met or that progress is made toward meeting the goals. In other words, Sustainable Seattle can measure progress toward achieving greater sustainability or the lack thereof, but it cannot directly change the City's policies and programs that affect this progress. Yet what makes the Seattle experience most impressive is the way that the City's leaders, particularly the City's administrative agencies, have begun to internalize the goals of sustainability.

In 1994, the City adopted its "Comprehensive Plan," called "Toward a Sustainable Seattle," that provides a statement of a "20-year policy plan that articulates a vision of how Seattle will grow in ways that sustain its citizens' values." This comprehensive plan represents a sustainability effort that is about as well-developed and coordinated as found in any U. S. city. The plan outlines policies that affect land use, transportation, housing, capital facilities, utilities, economic development, neighborhood development and planning, human development, and cultural resources in a fairly integrated way. A brief description of some of the major elements of this comprehensive plan provides a sense of the magnitude of this effort.

The Land Use element, for example, provides a full strategic plan for how to manage development in various kinds of urban villages within the city, where the plan specifies the kind of "village" each area of the city constitutes, and what kinds of land uses, employment, and housing densities are appropriate for each. Urban centers villages, hub urban villages, and residential urban villages each carry different functional land uses. An urban centers village has the densest land use, with high concentration of employment. Residential urban villages have the highest concentrations of low to moderate density residential development with a comparable mix of support services and employment. The Land Use element outlines numerous specific goals for each type of area, identifies the policies and legal authority governing the uses contained in each goal, and frequently sets specific targets and dates to be achieved. The established goals are driven, at least to some degree, by the overall goal of becoming more sustainable. For example, one of the land use goals is to "promote densities and mixes of use that support walking and use of public transportation" (City of Seattle 2000a).

The comprehensive plan also incorporates efforts to make the City government's internal operations more sustainable by fashioning something akin to a private sector company's internal environmental management system [EMS] (City of Seattle 2000b; Skinnarland 1999). Designed to comply with the international ISO 14001 environmental standards, this effort focuses on making all of the city government's internal operations consistent with sustainable practices. In Seattle, this EMS, referred to as an Environmental Management Program, was developed in two stages starting in 1996, where the first stage required all city operations and agencies to assess their environmental impacts and the legal requirements of the operations. With this assessment completed, a set of city-wide standards was established, and each city agency was required to prepare its own plan for how to implement these standards.

Each agency's plan had to address thirteen specific areas, including how it handles hazardous waste and hazardous chemicals, waste dumped on city-owned property, communication of workplace hazards to employees, the operation of petroleum storage tanks, energy and water conservation measures, waste reduction and recycling, environmentally responsible purchasing, the management of city fleet vehicles, and other areas.

The path-breaking sustainability efforts in Seattle have made the city both a model for other sustainable indicators initiatives, and a target of criticism. According to a recent study of over 170 state, regional, county, local, university, and ecosystem indicators projects conducted by *Redefining Progress*, a policy development organization in Oakland, California, at least 90 of these efforts apparently used Sustainable Seattle's indicators project as a model for their own projects (Sustainable Seattle 2000b). Yet the effort to pursue sustainability in Seattle has not been immune from criticism. As discussed previously in chapter 4, Seattle's land use policies, which flow from its comprehensive sustainability plan, have been attacked as unwarranted government regulation.

The Sustainable Seattle "model" or "approach" is one that prescribes the creation of a grassroots nonprofit organization that begins its initiative independent of city government or city agencies. The basic idea seems to be that once the organization takes hold, once it embarks on an indicators project and shows that it has the support of significant segments of the local population, then it can appeal to city policymakers— the mayor and city councilors—and city agency administrators to make the case that sustainability, as defined by the organization, should be on the city's agenda. To local advocates of sustainability, this model or approach seems to make perfect sense. Yet to at least one observer of the Seattle experience, the inability of the nonprofit organization to directly affect sustainability itself, constitutes a major shortcoming. Because the nonprofit organization possesses no legal authority for affecting public policy, and typically does not actually operate any pollution reduction programs directly, the impact of the organization on sustainability is indirect at best. This constitutes a major problem according to Brugmann, who has leveled serious criticisms of the Sustainable Seattle project. Brugmann acknowledges some important contributions of the project, but points to the limitations of the underlying model when he states that:

... Sustainable Seattle itself, organized as it was without connection to major institutions, generally, and the City's strategic and statutory planning processes, specifically, neither provided a blueprint nor stimulated commitments, nor even a consensus, for action. Its impact in driving change in local conditions was therefore, at best, catalytic. Reflecting this gap in the Sustainable Seattle approach, the City of Seattle established a quite separate city-wide task force in 1995 to develop a municipal strategic plan which is linked to the statutory 'Comprehensive Plan' required of all municipalities in the State of Washington. This task force identified its own long-term goals for Seattle. . . . While Sustainable Seattle set the stage for sustainability issues to be integrated into the Comprehensive Plan—and although there are thematic overlaps between the Sustainable Seattle indicators and the city's indicators—the municipality's indicators are not only different but, more importantly, are embedded in development policies that the municipality has a political and legal obligation to implement. (Brugmann 1997, 64)

Clearly, to Brugmann, the possible influence exerted by the Sustainable Seattle organization does not constitute a particularly important part of the process of moving toward sustainability. His view of the events of the last decade implies that there is little connection between the acknowledged successes of the organization and the decision by city policymakers to, in fact, move sustainability issues squarely onto the public agenda through its comprehensive strategic planning processes. When he begrudgingly refers to the effects of the organizations as being "at best, catalytic" Brugmann discounts the role that this organization may have played in helping to affect the local political agenda. Yet nothing in the mandatory strategic planning process in Seattle ensured that sustainability would become the cornerstone of the effort. Indeed, without the efforts of the Sustainable Seattle organization, it is entirely possible that sustainability would have played no more than a minor part in the city's planning.

Santa Monica One of the more aggressive sustainability programs in the United States is found in Santa Monica. Located in Southern California, just west of Los Angeles, Santa Monica is a city of about 90,000 in Los Angeles County, which has a population of about 9.8 million people. The population size of Santa Monica has been extremely stable since 1980, with only modest population growth from about

86,000 to its current level. Geographically, Santa Monica is a small city of only about 8.3 square miles, giving it a population density of about 10,800 people per square mile. The city has a full-time mayor and a city council of six members elected at-large on nonpartisan ballots.

One of the distinguishing characteristics of Santa Monica's sustainable city effort is that it was developed and is operated wholly within the city government itself (Brugmann 1997, 67–70). Santa Monica's sustainability efforts took initial shape in 1992, when the city's Task Force on the Environment developed its sustainability plan. This initial plan put forth a vision and a set of guiding principles and goals to help work toward achieving that vision. In September of 1994, the Santa Monica City Council adopted the plan as official policy, and this policy has been in effect since that time. The adopted Sustainability Program has sought to make the City more sustainable in four core areas: resource conservation; transportation; pollution prevention and public health protection; and community and economic development. The program included an indicators project, which produced a variety of specific ways of trying to measure how much progress the city was making toward achieving its vision in each of these four core areas. Additionally, the adopted Plan called for periodic assessment of this progress. Since that time, the city's task force has issued two "Progress Reports," one in December of 1996, and another in October of 1999.

The process of developing the Sustainable City Program in Santa Monica was a highly participatory one. The task force membership has varied since its inception, but generally has consisted of people who are associated with local environmental organizations, consulting companies, the academic community, and, sporadically, the business community. Although the vast majority of what is contained in the plan came from the work of the members of the task force itself, explicit efforts were made to engage a broad spectrum of people in the city. Public hearings and community workshops were held, meetings were held with numerous neighborhood associations and citizens groups, the Chamber of Commerce, and business and community leaders in order to solicit "input" for the creation of the indicators. There do not appear to be any formal records of exactly who participated, and in what ways the plan or any other activities of the task force actually incorporated information from these meetings and hearings.

Implementation of the plan is given to the Division of Environmental Programs in the city's Department of Environmental and Public Works Management. The Task Force on the Environment retains responsibility for providing "guidance" to the city agency. Because much of what is contained in the plan transcends what can be accomplished within a single agency, responsibility for implementing many of the component parts of the plan falls to administrative staff across many different functional departments. The city's Public Works Division has responsibility for water resource issues, the Transportation Department is responsible for public transit and related components, etc. In 1995, an interdepartmental Sustainable City Working Group, consisting of representatives from nearly all of the city's departments, was established. Among the many functions of this working group is the coordination of budget items in each represented department to reflect a common set of budget objectives.

A brief look at the city's indicators, baselines, and results over time, provides some sense as to the extent to which the city appears to take sustainability seriously. By far, most of the indicators and program elements focus on achieving improvements in the biophysical environment, although there are some quality of life indicators as well. According to the city's plan, every indicator that was selected had to meet the criteria that it (a) reflects something fundamental to the long-term economic, environmental, or social health of the community, (b) be measurable, i.e., either data already existed or a practical method of data collection could be created, and (c) represents something that could be influenced by community or governmental actions. For each selected indicator, whenever possible, a 1990 baseline was used as a starting point for assessing progress. Table 7.3 provides a summary of the indicators used in Santa Monica, along with their respective targets and achieved levels. Although there are no 1990 baselines reported in this table for many of the indicators, the City often does have measures for other intermediate years.

Resource conservation indicators focused on the amount of landfilled solid waste (in tons per year), with a target of a 50% reduction from the 1990 baseline of 124,000 tons. Other indicators include city-wide water usage (millions of gallons per year), energy usage (millions of mBTUs per year), and proportion of city office paper purchased that was

Table 7.3
Outline of Santa Monica's sustainability indicators

Indicator	1990 Baseline	Target for 2000	Achieved as of 1998
Resource conservation			
Landfilled solid waste (tons per year)	124,000	62,000	111,636
Waster Usage (million gallons per day)	14.3	11.4	12.4
Energy Usage (mBTUs per year)	6.45	Pending	Pending
Average postconsumer paper purchased (%)	22% (1993)	50%	Unknown
Transportation			
Annual ridership, municipal bus (million)	19.0	20.9	20.8
Average vehicle ridership, persons per car per day	1.13 (1993)	1.5	Pending
% City fleet vehicles using reduced emission fuel	10% (1993)	75%	33%
Pollution prevention/Public health protection			
Hazardous waste generated by City	Unknown	10% reduction	Pending
City purchases of hazardous materials	Unknown	10% reduction	Unknown
Underground storage tanks needing cleanup	25 (1993)	17	6
Percent Storage tanks meeting U.S. standards	43% (1997)	100%	95%
Percent Diversion/treatment of stormwater runoff	92% (1997)	100%	92%
Wastewater flows (million gallons per day)	10.4	8.8	9.05
Community and economic development			
Create and implement sustainable schools	No program	Full program	Proposal
Deed-restricted affordable housing units (public)	1,172	1,903	1,725
Public open space (acres)	164	180	180.6
Community gardens (number)	2	5	2
Trees in public spaces	28,000	31,263	29,263

Source: City of Santa Monica, Task Force on the Environment, "Sustainable City Progress Report Update," October 1999.

recycled. Transportation indicators measured the annual ridership on the Santa Monica Municipal Bus Line, the average vehicle ridership (persons per vehicle) for employers with over 50 employees, and the proportion of the city's fleet vehicles using reduced emission fuels. Pollution prevention and public health protection concentrated on reduction of city-wide use of hazardous materials, the number of underground storage tanks requiring cleanup, the amount of dry-weather stormdrain discharges to the ocean (in gallons average per day), and wastewater flows (in millions of gallons average per day). Finally, community and economic development indicators consisted of the creation and implementation of a sustainable schools program (the indicators is simply whether or not the program exists), the amount of deed-restricted affordable housing units, the number of acres of public open space, the number of community gardens, and the number of trees in public spaces.

Table 7.3 suggests how difficult it is for a city to define and implement a sustainability effort. For many of the indicators, there was no 1990 baseline information, and data for subsequent years had to be developed specifically for this project. In several cases, baseline information was still not available when the first progress report was written. Additionally, it is clear that accomplishing the target improvements is not always possible or easy. The amount of solid waste landfilled, for example, has been reduced from its 1990 baseline, but is still much higher than the established target.

It is perhaps ironic that one of the most important signs of how seriously Santa Monica appears to take sustainability comes from its own self-assessments and responses. According to the 1996 "Sustainable City Progress Report," only very modest improvements in the various indicators could be demonstrated (City of Santa Monica 1996, 1). Although many of the indicators revealed progress, the report noted that "Little or no progress towards meeting the indicator targets for Energy Usage, Ridership on Santa Monica Municipal Bus Lines, Community Gardens, and Implementation of a Sustainable Schools Program." Perhaps equally important, the self-assessment included a review of the processes used to implement the sustainability effort. As the report noted:

Despite progress made towards meeting the various indicator targets, sustainable policies and programs are still being undertaken on a "piecemeal" basis within the City. Coordinated implementation of the Sustainability Program within the

City has not yet been achieved. Many City staff are currently not aware of the program, and most of those who are aware of it do not see it as a high priority. ... The implementation plan ... has not been systematically carried out, staff responsibility for implementing the program has never been adequately defined, and staffing and funding necessary to properly implement the program have not been identified (City of Santa Monica 1996, 1).

The city's response to these issues was to redouble efforts at coordination, which included defining a new position called "Sustainable City Program Coordinator," whose job is to seek ways of improving interdepartmental and interprogram coordination. The result has been increased attention to staffing and coordination issues, with the development of specific employee performance evaluation guidelines that include sustainability elements, the creation of a "green purchasing group" consisting of purchasing agents from all city departments, and many other actions. The city's 1999 Report (City of Santa Monica 1999) reflected greater progress and optimism about achieving future results.

Boulder Boulder, Colorado, is a city with a population of about 95,000 in a county of about 272,000, in a metropolitan area (that includes Denver) of a little over 2 million, and is located just west of Denver. Its population size has grown considerably, having increased from about 67,000 people in 1970. It has a land area of about 25.4 square miles, giving it a population density of about 3,740 people per square mile. Boulder uses a Council–Manager form of government, with a city council composed of nine at-large members. The mayor and a deputy mayor are elected by the council from among its members.

The Boulder sustainability initiative involves a variety of city administrative agencies and the city council. The city council has organized itself into four committees, environmental sustainability, economic sustainability, transportation, and housing. The environmental sustainability committee proposed, and the full council approved, environmental priorities focused on pesticide reduction, waste reduction, increased energy efficiency, habitat preservation, and water quality. As part of these priorities, the council has approved purchase of hybrid vehicles for municipal government use, has enacted an ordinance requiring increased solid waste recycling, has embraced the use of an environmental management system for continuous review of the actions, programs, and policies of the city, and has considered an ordinance to implement the

Colorado Noxious Weed Act on the use of pesticides (Boulder City Council 2001). The city council has been aggressive about incorporating its priorities into the activities of the Boulder Valley Comprehensive Plan, the region-wide land use plan.

The involvement of the city's administrative agencies essentially consists of eight major components. These components focus on: (1) air quality; (2) sustainable internal city operations called the "Sustainability 2000 Project;" (3) a Greenpoints green building program; (4) a program for lawn care and integrated pest management; (5) a home energy check-up effort; (6) a comprehensive recycling program; (7) a Greenways program, and (8) a broad based collaborative effort with Boulder business called Partners for A Clean Environment, or PACE. This initiative is implemented largely by the Office of Environmental Affairs, which exercises considerable coordinating function between its programs and those of the Department of Public Works, the Office of Open Space, and the Department of Parks and Recreation. The city's efforts are overseen by the Environmental Advisory Board, made up of five members, appointed by the city council, who serve five-year terms. The city's efforts are also augmented by those of Boulder County, which include an aggressive Public Health Department that operates Air Quality, Solid Waste, Hazardous Waste, Pollution Prevention, and Vector Control Programs, and the Regional Air Quality Council, and the Boulder County Clean Air Consortium. Each of the city's seven components deserves a brief description.

The Office of Environmental Affairs operates the Air Quality initiative, and its mission focuses on efforts to reduce air pollution emissions at their point sources. A large part of its effort is focused on manufacturing industries located within the City. As the result of a voluntary reporting program initiated in 1996, Boulder manufacturing firms provided the city with reports on their efforts to reduce pollution to air, water, and land. The companies have developed overall pollution prevention plans and goals, and report annually on their progress. This program's purpose is twofold: (1) to provide clear, accessible information to the public about use and release of toxic or hazardous chemicals; and (2) to encourage businesses to develop specific pollution prevention goals and activities. Six companies—Ball Aerospace and Technologies, Lexmark International, Graphic Packaging, Hauser Chemical Research,

NeXstar Pharmaceuticals, and Roche Colorado—are the principal participants in this effort. The Air Quality initiative also works to raise residents' awareness of how they contribute to air pollution, with special attention to awareness about the contributions of auto emissions. The initiative sponsors a number of public information and education programs, including the "No Drive Day," the "Don't Top It Off" effort to affect how people fuel their cars, and many other programs.

The "Greenpoints Program" is an impressive effort to regulate the construction of residential buildings through the local building code. The idea behind this program is to create incentives for inclusion of environmentally friending building practices into the contruction of new homes. It requires contractors working on new building projects of between 501 and 2,500 square feet to "earn" at least 25 green points before they can be issued requisite building permits, and 1 additional point must be earned for each additional 200 square feet. Renovation projects are also required to earn points, but there are different targets. As part of the application for a city building permit, contractors must fill out a form identifying land use, framing, plumbing, electrical, insulation, heating, ventilation and air conditioning, solar, or indoor air quality provisions they have made, where each of a number of specific provisions earns a specified number of points. For example, a project whose design calls for using engineered lumber in the roof or floor would earn 5 points, where the entire project is required to earn at least 25 points in order to receive a building permit.

The Lawn Care and Integrated Pest Management program provides extensive advice to homeowners and landscape contractors concerning how to care for residential lawns without heavy use of fertilizers and pesticides or requiring excessive amounts of irrigation. It works to implement these lawn management techniques on the extensive lawn areas maintained by the City. It also operates a number of demonstration gardens to provide consumers with ideas on how to practice these techniques.

The Office of Environmental Affairs also operates a home energy checkup program designed to help residential consumers make their homes more energy efficient. It includes efforts to raise awareness among homeowners of their options to participate in either the WindSource or SolarSource programs operated by the local utility company. WindSource allows residential customers to participate in a program where they can elect to have a significant portion of their electricity come from wind-based sources. SolarSource provides consumers with the ability to install photovoltaic solar panels on their homes.

The Recycle Boulder program mainly involves curbside pickup of standard household recycled items, such as cans, newspapers, and plastics. It also maintains a yardwaste drop-off facility, and Boulder County operates a hazardous waste drop-off facility. Additionally, Boulder County has a nonprofit organization called "EcoCycle," a volunteer organization that is dedicated to promoting and conducting recycling activities.

The Boulder Greenways program was initiated as an outgrowth of the Boulder Creek Corridor Project, a plan for the preservation and development of Boulder Creek, developed in 1984. The project was designed to develop a diverse and aesthetically pleasing corridor along Boulder Creek that included a continuous off-street bicycle and pedestrian trail, to restore and enhance fish and wildlife habitat, and to develop passive park areas. The Program was actually established in 1987, when the City designated more than 20 miles of stream corridors along six tributaries of Boulder Creek as the Greenways Program. Funding for the Greenways Program was approved by City Council in December 1987. An original master plan, developed by the Planning, Public Works, Parks and Recreation and Real Estate and Open Space departments, was adopted by City Council in January 1989. A refined and updated master plan, design guidelines, capital improvement program and a detailed Greenways map were approved by Council in September 1990.

By far, the two most impressive components in Boulder's sustainability initiative are the city's internal sustainability efforts and the PACE program. The City's internal "Sustainability 2000 Project" is focused, as the project's subtitle suggests, on "getting our house in order." The city has developed ten key indicators that provide an overview picture of how much the local government's operations have themselves contributed to working toward sustainability, or at least toward being more environmentally responsible. These indicators refer to: (1) total water consumption by city government; (2) the amount of nonrenewable energy used by city facilities; (3) the proportion of the city's energy usage that is renewable; (4) the total amount of solid waste generated by the government; (5) the proportion of the city's solid waste that is recycled or composted; (6) the proportion of the city's purchases that are "environmentally preferred" (with minimal recyclable packaging, longer-lasting products, etc.); (7) the number of city-employee commuter trips; (8) the number of vehicle miles traveled for and during work by city employees; (9) the total amount of open space/mountain parks lands maintained by the city; and (10) the quality of the habitat (ecosystem health).

Each of these indicators carries with it a set of "action plans" and goals, as well as strategies to reach the goals. For a number of the indicators, the goal that is articulated is to develop mechanisms for measuring progress (for example, one goal related to materials management calls for the city to conduct an audit of actual trash volume produced by city facilities). According to the 1999 report of progress, the city had improved performance on five indicators (reduced water consumption, reduced use of nonrenewable energy, increases in city solid waste that is recycled, reduced vehicle miles traveled for work by city employees, and the total open space maintained by the city), and two indicators showed no change (the proportion of city energy used that is from renewable sources, and the total amount of solid waste generated by the city). One indicator showed decline (increased city employee commuter trips). And two indicators could not be measured yet (proportion of city purchases that are environmentally preferred, and the quality of the habitat).

Perhaps the most impressive element of Boulder's sustainability effort focuses on the relationship between major businesses and the rest of the city. In most cities, there is an iron wall between the public and private sectors, particularly when it comes to environmental issues. In Boulder, the Partners for A Clean Environment, or PACE, represents an initiative to get local businesses to voluntarily participate in making the city a more sustainable place (City of Boulder 2002). Started in 1998, this program consists of a cooperative effort between the city's Office of Environmental Affairs, the Boulder County Health Department, the Boulder Energy Conservation Center (a nonprofit organization), and the Boulder Chamber of Commerce. The focus of this program is to engage local small and medium-sized businesses in specific pollution prevention (P2) activities, with an aim toward reducing the use of hazardous materials.

There is a targeted outreach program to provide detailed information to specific businesses, and a certification and recognition program for those businesses that choose to participate.

Working in several specific business sectors, efforts have focused first on the auto repair, auto body, and printing sectors because of clear-cut pollution prevention alternatives. Additionally, the dental and restaurant sectors were targeted because of their impact on wastewater. These represent businesses that are usually exempt from the environmental regulations that apply to larger firms, but collectively they often still create significant environmental impacts. In each sector, PACE staff held focus group interviews with local business representatives to develop sets of achievable and appropriate pollution prevention goals for that sector. These goals then became the criteria for business certification. In other words, once the goals were specified, any specific business in the sector could become certified as a "partner" by meeting all of these goals. A partial certification was instituted for businesses that could not meet all of the goals but could meet at least three of them. In addition to the PACE certification effort, there is also an effort to get local businesses, particularly manufacturing firms, to report on their own hazardous materials usage and generation, and their internal efforts to reduce their streams of hazardous wastes.

Portland Although Portland, Oregon, is not often identified as a prototypical sustainable city, there is little question that upon examination it possesses one of the most impressive sustainability initiatives of any major city. Portland has a population of about 512,000 in a metropolitan area of about 1.8 million people. Portland uses a commission form of government, where the mayor, four commissioners, and the auditor comprise the city's six elected officials. Once elected, each commissioner becomes the head of an administrative agency (one commissioner, for example, serves as the head of the Department of Public Safety). A single commissioner heads the administrative offices that have principal responsibility for environmental and related programs. All commissioners are elected at-large on a nonpartisan basis and serve four-year staggered terms.

Portland was one of the earliest cities to become involved in explicit efforts to become more sustainable, enacting in 1993 a policy on global warming that called for the reduction of the city's carbon dioxide emissions by 20% from 1990 to 2010. In 1994, the city adopted a set of ten sustainable city principles that reflected a long-term commitment of the city government to pursue a variety of specific policies. Portland's sustainability initiative is perhaps most like that found in Seattle where sustainability goals are an integral part of the city's Comprehensive Plan. The role of sustainability in the city oozes out of every ounce of the city's government operations, and affects the way the government is organized and functions. The city established a single agency with central programmatic and coordinating functions, the Office of Sustainable Development, formerly called the Office of Energy. In conjunction with the Sustainable Portland Commission, an organization of volunteers appointed by the mayor, this office operates or coordinates programs on business conservation, residential conservation, global warming, solid waste and recycling, and sustainable development. Much of the clean water operations and services fall to the Bureau of Environmental Services, headed by the same commissioner who heads the Office of Sustainable Development.

Among the many elements that make up the city's sustainability initiative is the creation of "sustainability benchmarks," which are essentially the same as sustainable indicators in other cities. The difference in Portland is that the benchmarks are designed to explicitly compare how the city is doing to a select group of other cities, including Charlotte, Denver, Kansas City, Sacramento, Seattle, Austin, Minneapolis/St. Paul, and Phoenix (Sustainable Portland Commission 2000). The latest report suggests that Portland has fared better than the average for the other cities in its air quality, its carbon dioxide emissions, maintaining a "tree canopy," and the rate of residential solid waste recycling. It fared below average in vehicle miles traveled and traffic congestion, and releases of toxic chemicals.

One mechanism that Portland has used to promote sustainability is in its use of zoning and land use regulations. It has adopted an approach to zoning that fairly aggressively takes the environmental protection and the potential for detrimental environmental impacts into consideration in its regulation of land use. To do this, Portland explicitly incorporates an "environmental overlay" on its zoning and land use plan. This overlay is used to designate special areas of the city that need to be treated with greater sensitivity to environmental impacts. As discussed briefly in chapter 4, and shown in figure 4.1, regardless of how an area may be zoned, a designated area within or across zones could be defined as an environmental zone or as a transition area around an environmental zone, and these areas receive special protection. Of course, as discussed in chapter 4, the aggressive use of zoning to control growth in this manner has been subjected to great criticism from those who believe that such controls usurp private property rights, and from those who believe that the pursuit of economic efficiency should take priority over the pursuit of sustainability (Charles 1998).

Portland also operates a Green Building Initiative, an ambitious program to ensure that city infrastructure and capital improvement projects take advantage of a green building rating system, and to provide technical assistance to the building and construction industry. Although this program does not go as far as that of many other cities that have established green building standards as a prerequisite for obtaining building permits, it nonetheless provides a clear and well-established set of standards (the Leadership in Energy and Environmental Design rating system developed by the U.S. Green Building Council) for what constitutes a green building.

Perhaps the most important part of Portland's sustainability effort can be found in its Comprehensive Plan. Like the Comprehensive Plan found in Seattle, sustainability represents a high priority that permeates Portland's Plan. Although the environment and energy constitute two of the twelve main elements in the plan, sustainability goals are threaded throughout the plan. The plan specifies for each element, including energy and the environment, a set of goals along with the policies, programs, and actions that are necessary to achieve those goals. In 1996, for example, the plan called for the city to "promote a sustainable energy future by increasing energy efficiency in all sectors of the city by ten percent by the year 2000" (City of Portland 1999). The goals are accompanied by specification of numerous specific objectives, along with two-year and long-term action plans for accomplishing these goals and objectives. For example, a two-year action plan for a goal of increasing waste reduction and recycling specified that the city must "set up recycling programs for an additional 500 multifamily buildings and 20 downtown commercial buildings." Frequently, the Plan specifies which agencies are responsible for accomplishing the goals, and identify any changes in policies or ordinances that may be required to achieve them.

San Francisco   San Francisco is the largest of the eight cities profiled here, with a current population of a little over 800,000 people, located in a Bay Area with a total population of over 6.6 million. The population of the city has grown modestly but steadily over the last twenty years, increasing from about 678,000 in 1980 to almost 724,000 by 1990. Geographically, San Francisco is a compact city with land area of only about 46.7 square miles, giving it by far the densest population among the eight cities of over 17,000 people per square mile. The city is governed by an eleven-member Board of Supervisors, elected by districts, and an independently elected mayor.

San Francisco's sustainability efforts are notable for the remarkable breadth of its conception of sustainability and for its lack of success in moving sustainability issues onto the public agenda. Sustainability in San Francisco received a major push in 1996 with the release of its Sustainability Plan (City of San Francisco 1996). This plan created the blueprint for a wide array of programs, policies, and resolutions across numerous substantive areas, including air quality, biodiversity, energy, climate change and ozone depletion, food and agriculture, hazardous materials, human health, parks, open spaces and streetscapes, solid waste, transportation, water and wastewater, the economy and economic development, environmental justice, and several other areas.

The 1996 Sustainability Plan for the City of San Francisco, officially adopted in early 1997, was developed beginning in 1994, and involving hundreds of people, numerous neighborhood meetings and public hearings, reflects as broad an interpretation of what constitutes sustainability as that found in any city (City of San Francisco n.d.). The plan includes an indicators project that provides measures of progress in each of the substantive areas listed above. Indicators for air quality focus on the number of existing buildings in the city that have joined the Building Air Quality Alliance voluntary program, the number of people reporting to health providers with respiratory problems, and the percentage of new cars which are low, ultra-low, or zero-emission vehicles. Direct measures of are quality, such as the ambient pollution measures, and not included as indicators. Biodiversity indicators include the number of native plant species planted in developed parks, private landscapes, and natural areas, the count of the number of different bird species sighted, and others. In short, the San Francisco indicators project covers a rather diverse array of areas.

The sustainability initiative, particularly the indicators project, was the product of the operation of a nonprofit organization called "Sustainable City," which not only sought to develop the indicators but also to serve as an advocacy voice to help place sustainability on the city's public agenda. Largely as a result of the work of this organization, the city government's involvement has been through the city's Commission on the Environment, consisting of seven members all appointed by the mayor, which is charged with setting policy and advising all other city agencies and the Board of Supervisors. The commission works primarily with the city's Department of the Environment, which is the central agency with any responsibility for developing and implementing the sustainability plan itself. The Department of the Environment is one of a handful of departments in the city that reports directly to the city's professional chief administrator.

Finding programmatic manifestations of the sustainability program in the city's administrative agencies is somewhat more challenging than in most other cities profiled here. The Commission on the Environment and the Department of the Environment are fairly active in identifying specific and narrowly targeted issues to confront and to seek approval from the Board of Supervisors, and the Department has focused much of its attention on ordinances that require, and pilot projects that achieve, greater resource efficiency in city buildings. The standard operating mode for the commission is to present specific issue resolutions to the Board of Supervisors for its ratification. Over the last several years, the Board has approved resolutions on such issues as discouraging the purchase of nonrecycled beverage bottles; the intentional release of balloons into the air; encouraging the use of alternatively fueled public transit vehicles; pesticide reduction; and numerous other issues (San Francisco Commission on the Environment 2002).

Despite this, there is little evidence that the Department functions within any sort of comprehensive sustainability plan. Moreover, there is reason to believe that the pursuit of sustainability is constrained by high-profile opposition. In late 1998, lamenting the lack of political support for the ambitious initiative, the President of Sustainable City, the nonprofit organization, presented his assessment of San Francisco's experiences at the California Community Indicators Conference (Redefining Progress 1999). This assessment outlined a number of impediments in moving from an indicators project to establishing programs and policies to make sustainability happen. His observation was that:

Neither the current president of the Board of Supervisors nor the mayor have [sic] bought into the need for strategic sustainability planning and implementation of the sustainability plan. This has resulted not only in the non-funding of the sustainability mandates of the new Department of the Environment, but several threats to the effectiveness and existence of existing city environmental programs. . . . The San Francisco Chronicle, the city's major newspaper, is very conservative on environmental issues. Their dismissive, ridiculing coverage of the sustainability plan upon its introduction at the Board of Supervisors caused many elected officials to shy away from associating themselves with it. (Magilavy 1998, 3)

So, while San Francisco presents an impressive conception of sustainability, as reflected in its Sustainability Plan, in terms of commitment by the city government, the city would seem to take sustainability somewhat less seriously than other cities profiled here.

Jacksonville   Jacksonville is a fairly large, rapidly growing sunbelt city located in the northeast corner of Florida. Its current population stands at over three-quarters of a million people, having grown over 35% in twenty years, from 571,000 in 1980 to 672,000 by 1990, to its current population. The "urban core" of the city has declined in population by about 5% over the last ten years, while the growth has taken place largely in the outer rings of the city. Due to the fact that Jacksonville underwent city–county consolidation several decades ago, the city is coterminous with Duval county. In land area, Jacksonville is the largest of the eight cities profiled here, composed of 840 square miles. This gives Jacksonville by far the lowest population density, with about 920 people per square mile. The city is governed with a strong mayor–council structure by a city council of nineteen members, five elected at-large and fourteen elected by districts. The mayor is elected independently of the city council.

Jacksonville represents a city that has many pieces in place that would appear to suggest that it takes sustainability reasonably seriously. Yet there is little coordinated action taking place within the city government itself, and there is something of a disconnect between the organizations working on livability issues and most of the city's administrative agencies. Jacksonville's sustainability initiative began in earnest in 1985, and has largely been conducted as its "Quality of Life in Jacksonville" project. This project represents a collaborative effort by the City of Jacksonville, the Jacksonville Chamber of Commerce, and a nonprofit organization called Jacksonville Community Council, Inc. A substantial portion of Jacksonville sustainability effort is contained in the formal "indicators of progress" adopted to measure the extent to which the city is successfully improving the quality of life (City of Jacksonville 2000). Indeed, Jacksonville has certainly been a national leader in the adoption of such indicators, and that fact was apparently very influential in the early activities of Sustainable Seattle (AtKisson 1996). Through a voluntary process that involved about 140 people and funding from the city government, by 1991 this project had established explicit indicators and goals in its "Target for 2000." In 2000 and 2001, the process was repeated through the participation of residents in designated task forces (some of which continued to exist from the earlier effort) in order to produce an "indicators upgrade project." In this project, the goals established earlier were assessed, indicators were revised, and new goals were established in its "Target for 2005."

The heart of the Quality of Life project is in the indicators themselves. The indicators are organized in nine different "elements" or categories: public education; local and regional economy; natural environment; social environment; culture and recreation; health; government and politics; mobility (transporation); and public safety. Each of these elements contains numerous indicators (eighty-one in the current indicators project), many of which are derived for an annual survey of city residents sponsored by the Council. A single indicator was designated as a key to each element. For example, the key indicator for public education is the percentage of students entering high school who graduate in four years. The target for 2000 set in 1991 was 90%, which would have represented a significant improvement over its 1990–91 graduation rate of 76.1%. Over the last five years, the graduation rate experienced a precipitous decline, reaching 58.7% for the 1998–99 year. This, of course, is illustrative of the limitations of an indicators project that is not explicitly linked to a strategic plan or action plan of some sort. Observing that the indicator has moved in the wrong direction provides no guidance or prescription for reversing the trend.

The 2000 Progress Report (City of Jacksonville 2000) does not paint a rosy picture of success in improving the quality of life. Of the nine key indicators, two moved in the wrong direction (things got worse in public education and length of time to commute to work), two fell short of their goals with virtually no progress toward improvement (perceptions of racism and infant mortality), one fell short but with movement in the right direction (things got a little better in growth of new employment), three met or exceeded their goals (things got better in air quality, perceptions of the quality of public officials' leadership, and crime rates), and one was not measured (the number of entertainment events a major public facilities).

On its surface, the major shortcoming of the Jacksonville sustainability effort is in the lack of a direct connection between the indicators project (and the indicators themselves), and the prescribed actions necessary to accomplish the state goals. In other words, the indicators project itself does not include any sort of action plan that might prescribe the policies, programs, or activities that would be necessary to meet the goals. However, that does not necessarily mean that there is no action plan or that there is no link between the indicators project and the agenda of city government. Indeed, among the task forces created in the context of the indicators project were several "Implementation Task Forces" that have directly taken on this challenge. For the most part, however, the actions prescribed have not been in the form of specific policies or programs; rather they have been oriented around public educational and advocacy initiatives designed to heighten awareness of the problems of the city and what some of their solutions might be (Besleme, Maser, and Silverstein 1999, 17). Additionally, the Jacksonville Community Council, Inc., a nonprofit organization (funded by the city and by the United Way) and the Chamber of Commerce have organized initiatives to help work toward achieving results in several of the indicators areas (Besleme, Maser, and Silverstein 1999, 20).

The fact remains that the livability and sustainability efforts in Jacksonville, with the possible exception of some elements of the city's recently completed comprehensive plan, described later, occur largely outside of the realm of the city's government. The indicators project has set the stage for serious progress toward sustainability, but the question remains whether the city's government has been able to act on that stage. There are certainly a number of programs and policies that might be said to indicate that the city does take sustainability seriously. For example, in 1998, Jacksonville was designated an EPA brownfield pilot project city, and some progress has been made to identify and remediate the brownfield sites that make them productive for the city's residents. The city's Regulatory and Environmental Services Department operates a mobile air toxics monitoring laboratory that moves around the city for two weeks at a time to measure and record ambient air quality. Transit policies are geared toward encouraging greater carpooling and less reliance on the automobile. And there is an effort to increase the amount of open space areas of the city. At the same time, there is a notable absence of policies and programs in some key areas. For example, there is virtually no attention given to energy issues and energy conservation, or to how the building code might be modified to make new construction more consistent with sustainability goals.

One of the more telling ways to ascertain the role of sustainability on the Jacksonville public agenda is to examine the city's Comprehensive Plan. In 2000 and 2001, the city developed its ten-year Comprehensive Plan. Divided into thirteen elements focused on historic preservation, housing, mass transit, port and aviation facilities, traffic circulation, recreation and open space, public utilities, conservation/coastal management, future land use, and several other areas, the plan does not appear to take any sort of integrated approach to sustainability or livability. It is clear that issues related to sustainability are threaded through several of the plan's elements, particularly the public utilities and future land uses elements. In the former, much attention is given to protecting and conserving water supplies, and to management of solid and hazardous waste. Although the plan calls for movement toward integrated solid waste management, it is also clear that being aggressive in achieving higher levels of recycling is not a particularly high priority. Language that specifies policies that will achieve results "within existing funding levels" and "in an economically practical manner" may hint at the lack of such commitment. Certainly no one advocates that cities should act in economically impractical manners, but when these terms are used as code words to suggest that they are not sufficiently high priorities to receive attention in the city's operating budget, this can hardly be seen as consistent with efforts that take sustainability seriously.

The future land use element of the city's comprehensive plan probably provides the most significant evidence that the city has some serious interest in sustainability. This plan element calls for the use of zoning to define urban villages with mixed uses to minimize automobile reliance, the delineation of industrial areas that encourage location of manufacturing industries in areas that are less environmentally sensitive, and other efforts to minimize sprawl. This is particularly important for a city the size of Jacksonville. If the city aggressively applies the land use plan, significant improvements in achieving sustainability goals could be accomplished.

Basic Commonalities and Differences among the Cities

The descriptions of the eight cities provide abundant evidence that, despite sharing a concern for sustainability and the environment, the details concerning what this means vary considerably from place to place. The content of cities' sustainability efforts looks very different in Chattanooga than it looks in Seattle. However, the fact that these cities do share concern for sustainability raises the question of whether they share other characteristics as well. As already discussed, these eight cities represent a range of population sizes and recent experiences with population growth or decline. Are there other basic characteristics shared by these cities, particularly characteristics that might help explain why these cities have pursued sustainability when others have not? Or are these cities so different from one another as to make explanation problematic? No effort is made to compare these eight cities to cities that have not embarked on sustainability initiatives. And a more thorough comparison of all 24 cities identified in chapter 2 is conducted in the next chapter. Here the question simply is whether there seem to be commonalities among the eight profiled cities.

The basic information in table 7.1 provides a thumbnail sketch of the eight cities in terms of their demographics. Clearly, there are some similarities, but the data also reveal major differences. The cities represent a range of population sizes, from the relatively small cities of Santa

Monica and Boulder, to the large cities of San Francisco and Jacksonville. Nearly all of the cities experienced considerable population growth since 1980 and in the decennial period just prior to the advent of their sustainable cities initiatives. Only Chattanooga lost population, dropping in size by nearly 13% over a twenty-year period. One city, Austin, nearly doubled in size from 1980 to 2000, while Jacksonville and Portland grew substantially as well, although much of Portland's population growth over this period was the result of annexation. These cities also represent a wide range of population density, ranging from San Francisco's over 17,000 people per square mile to Jacksonville's less than 1,000 people per square mile.

In terms of major similarities and differences among these cities' sustainability initiatives, there are several. First, all of the cities except Chattanooga have an aggressive indicators project, and most began the process of initiating sustainability with that project. There are significant differences among the indicators projects—in Jacksonville the preference is to call them quality of life indicators, although substantively there is little difference between these and other cities' sustainability indicators. Second, in nearly every city, the sustainability initiative either got started or received a significant boost from a nonprofit, nongovernmental organization. Whether it was the Jacksonville Community Council, Sustainable City (San Francisco), or Sustainable Seattle, among others, these nonprofit organizations played a significant role in the early development of sustainability initiatives.

There certainly are no data to bring directly to bear on the question of whether or to what extent there may be a relationship between how communitarian a place the city is and how seriously it takes sustainability. Indeed, there are very few ways that broad-brush political characteristics of the cities can be associated with the sustainability. However, one issue that arises in such an assessment concerns the relationship between the local political ideology of the city and a willingness to address sustainability problems. Is sustainability really the product of a particular political ideology? If so, is it merely a product of a liberal political establishment in the city? There are no available data that would permit an assessment of this possibility directly. As an alternative, admittedly crude, proxy, the question can be posed concerning whether there is a relationship between the political party identification of the city's

Table 7.4
Presidential vote results in eight sustainable cities

City	2000 percent Gore/Lieberman vote	2000 percent Bush/Cheney vote	1996 percent Clinton/Gore vote	1996 percent Dole/Kemp vote	Average percent Democrat vote	Average percent Republican vote
Chattanooga	43.0	55.3	43.3	49.8	43.2	52.6
Jacksonville	41.1	58.5	44.4	50.1	42.8	54.3
Austin	41.7	47.0	52.3	39.9	47.0	43.5
Boulder	50.1	36.4	50.7	33.5	50.4	35.0
Santa Monica	63.4	32.4	53.0	39.0	58.2	35.7
Seattle	60.0	24.4	57.0	31.8	58.5	28.1
Portland	62.9	27.9	58.4	26.0	60.7	27.0
San Francisco	74.4	15.9	72.2	15.7	73.3	15.8
Average	*54.6*	*37.2*	*53.9*	*35.7*	*54.2*	*36.5*

residents. To examine this, the last two Presidential elections' results from the eight cities were compiled, and are presented in table 7.4. The table shows that while some of the cities tend to be very heavily Democratic in their voting, others are less so. San Francisco, Portland, Seattle, and Santa Monica are pretty Democratic kinds of places, at least based on how they voted in the last two Presidential elections. On the other hand, Jacksonville and Chattanooga are primarily Republican kinds of places. And Austin and Boulder fall in the middle. Although this does not provide definitive answers, it is clear that being a Democratic place does not seem to serve as a prerequisite for taking sustainability seriously.

Clearly, the eight cities examined in this chapter represent a wide range of basic demographic and political characteristics. Although the eight cities share a high degree of interest in sustainability, they share few other characteristics. They represent a very diverse set of cities. They are not all very large, or very small, in population size or land area. They are not all highly dense in population, nor are they all low-density, sprawling kinds of places. They are not all wealthy, and they are not all poor. Not all have large minority populations, although some certainly do. And they are not all politically similar—some tend to vote Democrat and some Republican. In short, this group of cities, all of which consider sustainability to be important, is a very diverse group of places. Whether the full array of twenty-four cities examined in chapter 2 might be said to exhibit patterns based on how seriously they seem to take sustainability is an issue examined more thoroughly in chapter 8

8

Sustainable Cities in Practice: More Cities, More Questions

If the eight cities profiled in chapter 7 represent examples of the best practices among U.S. cities, then how seriously do other cities seem to be taking sustainability? Are there other cities that are also moving in the right direction even if they may not have achieved quite as tangible results as these eight cities? Are there other cities that seem to have done just as much as those profiled previously? As noted earlier, the eight cities are, in fact, not the only cities working on sustainability, and indeed there are certainly other candidates that could have been profiled. But that still leaves open the question about what other cities are doing. One purpose of the chapter is to review some of unique features of some of those other cities, and to provide a sense of how they measure up, in a less than perfectly systematic way, to the eight cities. Moreover, this chapter uses the information about the eight profiled cities along with other cities as a backdrop to raise a number of questions and hypotheses that can be addressed through comparative empirical analysis, and questions that still need to be addressed or hypotheses that warrant future testing. If the analysis in this book represents a preliminary inquiry, it would be greatly remiss if it did not, at least in part, serve the purpose of providing some conceptual guidance for future research and analysis. It does this by outlining some of the central research questions that emerge either from the existing literature on sustainability, or from the experiences of specific cities. This discussion is accompanied by an effort to suggest some research approaches that would help to address the outlined questions.

Sustainability in Additional Cities

Although the eight cities profiled in chapter 7 represent a range of the more elaborate, ambitious, and perhaps successful, sustainable cities programs in the United States—cities that do appear to take sustainability seriously—there are additional cities that have made efforts to move in that direction. To be sure, there are other North American cities whose sustainability efforts are every bit as elaborate as these eight, including Toronto and Scottsdale, and perhaps San Jose and Tampa. But there are also a number of U.S. cities whose initiatives might be said to be fledgling at least in the sense that they have not achieved the programmatic breadth or depth of others. Cities such as Olympia, Washington; Boston, Cambridge, and Brookline, Massachusetts; Cleveland, Ohio; New Haven, Connecticut; Denver, Colorado; and Santa Barbara, California, are among these cities. It may not be possible here to know whether these cities' initiatives are less elaborate because they simply do not take sustainability as seriously, or because they are at an earlier stage in their evolution.

All of the eight cities profiled earlier went through the sometimes-lengthy process of developing their initiatives, and had they been examined three, four, or five years ago, many of them would have seemed far less serious. As we shall see, in some cities, such as Olympia, Washington, and Cambridge, Massachusetts, work on sustainability issues has been under way for some time, and given that, one might expect greater progress. On the other hand, Brookline, Massachusetts, and Tampa, Florida, have started their initiatives relatively recently, and consequently great progress might not yet be expected. Of course, it is not possible to tell with any confidence whether the cities whose initiatives are of relatively more recent vintage will make as much progress as some other cities, but they do provide abundant fodder for speculation. The purpose of the descriptions that follow is to provide a succinct overview of some additional cities' sustainability efforts. These are provided as a comparison and contrast to the earlier, more in-depth descriptions.

Olympia, Washington's sustainability effort stands in some contrast to that in Seattle. Although both cities operate under the same state-mandated, comprehensive planning framework, there are significant differences between the cities. Like Seattle, the initiative for sustainability in Olympia essentially originated with a nonprofit organization. The nonprofit organization Sustainability Roundtable, Olympia's counterpart to Sustainable Seattle, has done the lion's share of work on the issue. Olympia is a much smaller city than Seattle, with a 2000 population of about 42,000, but its sustainability initiative saw its germination in 1991. This initiative successfully developed an indicators project, but the early efforts did not facilitate a particularly close connection with city government. The city adopted sustainability as a guiding principle in 1995, but there is little evidence that the operation of the Roundtable played much of a role in this. Despite the fact that the city council gave the indicators project a substantial boost with a small grant, the city and the Roundtable never developed any sort of working relationship. At least one activist in the Roundtable suggested that the city did not understand sustainability, and in implementation "... they're doing it in bits and pieces, so the actual meaning of true sustainability gets lost somewhere in the rhetoric" (Craig 2000). In addition to the indicators project of the Roundtable, the city government itself (the department of public works) has developed a set of sustainability indicators that apparently bear little relationship to that developed by the Roundtable. The Roundtable also operates a program called "Sound Exchange" in which consumers can purchase "Sound Hours," exchangeable for either one hour's worth of work or $10 worth of goods, from local merchants. It represents an effort to foster the idea of keeping local money within the city and sustaining the local economy.

Boston, Brookline, and Cambridge, Massachusetts—three cities in close geographic proximity—have all made some progress toward creating sustainability programs. In Cambridge, a city with a year 2000 population of just over 100,000 (up from 95,800 in 1990) and a very high (15,000 person per square mile) gross population density, the initiative was spearheaded by the Cambridge Civic Forum, a nonprofit organization that promotes active community participation in discussions about the city's future, and the Cambridge Coalition for Sustainability, a local citizens group. These groups initiated an indicators project in the early and mid-1990s, and a 1993 report entitled "Towards a Sustainable Future" (City of Cambridge 1993). Largely because of little responsiveness from the city government, this initiative produced little in the way of results. There is certainly little evidence that the Forum had much impact on the city government itself, and how it deals with environmental issues. Within the city's administrative structure, environmental issues tend to be scattered and fragmented. Most of the city's environmental activities and programs are operated out of its Community Development department, including transportation planning, lead abatement programs, coordination with other agencies working on water and air quality issues, and an aggressive bicycle lane project. These various activities are not a central part of any sort of comprehensive sustainability plan or operation, and there is no official indicators project that monitors progress or guides administrative decisions. In the last couple of years, the city government, particularly the Department of Community Development, has revisited issues of and started planning for, sustainability. This has manifested itself in the activities of the Cambridge Growth Management Advisory Committee, and a variety of new zoning ordinances, passed in 2000, to affect some additional control over development (City of Cambridge 2001b). Although the Cambridge city government operates a number of specific programs that are consistent with goals of sustainability, such as an aggressive curbside recycling program, and a comprehensive pedestrian plan, there is no centralized mechanism for administering these programs.

In Boston, the city established its Sustainable Boston initiative in 1996 and it quickly took a back seat to traditional economic development in the city's priorities (City of Boston 2000). Boston is a city of well over half a million people, and a high population density of nearly 13,000 people per square mile. While the city operates a Department of Environment, which has responsibility for the current city-sponsored sustainability initiative, most of its activities are oriented around providing a standard package of inspectional and related services, and compliance with federal and state environmental regulations. The city, working with the nonprofit Boston Foundation, has made an effort to develop an indicators project represented by publication of *The Wisdom of our Choices: Boston's Indicators of Progress, Change, and Sustainability*, a report released by the Foundation and the City of Boston at a Boston Citizen Seminar in October, 2000 (Boston Foundation 2000). The indicators project report focuses on ten functional areas or "sectors," providing measures of performance over approximately the previous ten years. The ten areas include civic health, cultural life and the arts, the economy, education, the environment, housing, public health, public safety, technology, and transportation.

The fate of this city's sustainability initiative symbolizes the tension with economic development, illustrates the difficulty of getting sustainability on the public agenda when there is little support from top political leadership, and raises the question of where, within a city's administrative organization, a serious sustainability initiative belongs. As an official from the Boston Redevelopment Authority (BRA), the city's chief arm for economic development, stated, "right now, I don't think sustainability has registered on a plane where people understand it. I don't think that people are formally rejecting it or embracing it . . . it's just waiting for something to happen. . . . The only way something will happen is if the mayor makes it a priority, and he hasn't done that" (BRA 2000). The same representative argued that the BRA was the agency that should have responsibility for the sustainability program rather than the Environmental Services department. In his view, only the chief planner has the authority to make all the connections among programs and issues that sustainability requires. As he stated it, ". . . this program needs to be with the chief planner in the BRA. If it's stuck in the environment department, and someone from the department calls, say, the school department about an indicator or its performance, they're going to say 'Come on, will ya, we have enough to do here'" (BRA 2000). In summary, the city of Boston has made some progress toward establishing a sustainability program, and its accomplishments to date have largely been focused on an indicators project. Future progress will depend on the city's ability to define a comprehensive set of action plans—specific programs and policies—oriented toward achieving tangible and coordinated results.

Brookline, a town of just over 50,000 people and population density of about 8,000 people per square mile that is governed by a representative town-meeting form of government, was slow to take up the issue of sustainability, but has now done so. Spearheaded by its Partnership for Sustainable Brookline Project, the city's Planning and Community Development department has now made efforts to begin building sustainability into its ten-year comprehensive planning process. The draft plan, as of 2001, contained elements focusing on parks and open space, water quality, sustainability, historic preservation, commercial development, housing, schools, recreation, transportation, town facilities, and telecommunications (Town of Brookline 2001a). Within the sustainability element, the focus is on monitoring greenhouse gas emissions, energy usage, solid waste management, transportation emissions, and local smart growth with an eye toward improving efficiency in all of these areas (Town of Brookline 2001b).

Cleveland is a fairly large city, and despite having lost over 16% of its population from 1980 to 2000, the population currently stands at just under half a million. The sustainability initiative in Cleveland is composed of a number of projects and activities, some located in the nonprofit sector and some in the Cuyahoga County Planning Commission. Few, if any, of these activities are located in the city's administrative agencies. The nonprofit sector activities include the city's Ecocity Cleveland project, the Sustainable Cleveland Partnership, and some of the activities of the Cleveland Neighborhood Development Corporation. They also include a variety of environmental monitoring programs operated under the auspices of the Northeast Ohio Environmental Monitoring for Public Access and Community Tracking (NEO-EMPACT) program, an EPA-supported coalitional partnership among numerous organizations in that part of the state (NEO-EMPACT 2001). NEO-EMPACT engages in several technically oriented efforts to provide guidance and advice on air quality and on monitoring and assessing urban sprawl. For example, it provides real-time air quality monitoring for the region of Ohio, and makes available a variety of technical "tools," including Geographic Information System (GIS) materials, to help in the assessment of growth and growth management strategies.

Another part of the Cleveland sustainability story is found in its Ecocity Cleveland (Ecocity Cleveland 2001). The Ecocity Cleveland initiative includes specific ecovillage projects, particularly the Cleveland EcoVillage (2001). The idea behind this ecovillage project was to identify a section of the city that desperately needed redevelopment, and to design this redevelopment from the bottom up with an eye toward making it as environmentally friendly as possible. Actual construction began or portions of the development in 2000, with additional projects getting under way thereafter (Beach 2000).

The Cuyahoga County Planning Commission has developed a variety of sustainability-related initiatives, particularly in the area of regional public works integration, greenspace planning, brownfields redevelopment, and land use impact assessment. For the most part, however, the Planning Commission focuses on providing information and technical assistance to cities and towns, and does not operate specific sustainability programs of its own.

Tampa and Hillsborough County, Florida, have taken a somewhat unique approach to its sustainability initiative. Tampa is a city of nearly 300,000 residents in a county of about one million people. In many respects, the Tampa–Hillsborough County effort deserves to be discussed in the context of cities that take sustainability fairly seriously. Once the project transitions beyond the demonstration stage, it may well compare favorably with programs discussed in chapter 7. It is unique because of the unusually clear role played by the state government, and because it is designed from the outset to require an unusual amount of coordination with the county government. As described elsewhere, many sustainable cities projects make efforts to coordinate their activities with their surrounding counties, but these efforts are typically informal and experience only limited success. In Tampa, the city–county arrangement is more formal. Additionally, perhaps more consistent with prescriptions for a process driven by technical competence within city government, Tampa has enlisted the assistance of a consulting firm with expertise in planning for sustainability.

The Tampa–Hillsborough County effort has been designated as one of several sustainability demonstration projects created by the Florida state Department of Community Affairs under a 1996 state-wide statute (State of Florida 2001). The Tampa project became official when the city and county entered into a formal agreement with the state agency. The agreement calls for the city and county to pursue six principles of sustainability: "restoring key ecosystems; achieving a more clean, healthy environment; limiting urban sprawl; protecting wildlife and natural areas; advancing the efficient use of land and other resources; and creating quality communities and jobs" (City of Tampa 2001). By virtue of entering into this agreement with the state, under the governing statute the city and county are provided with greater legal authority to approve development proposals and plan amendments without state approval. Since that time, Tampa has put into place a number of planning initiatives primarily oriented toward economic revitalization, brownfield redevelopment, infill and neighborhood revitalization, public transportation, and a variety of specific projects.

Working in conjunction with the supporting state agency and four other cities' demonstration projects, the Tampa project developed a set of community indicators covering a number of functional areas: economic development and trends in development; affordable housing; the natural environment and natural resource conservation; economic development; quality of life; disaster preparedness; and transportation. The second annual report for this project, issued in mid-1999, suggested that additional indicators would be forthcoming (City of Tampa 1999). The process of developing the sustainability initiative in Tampa has largely been a top-down one, where the city and county planning departments have played the dominant roles. Participation by others in this demonstration project has been limited to that from the specially created Sustainable Communities Advisory Board, consisting of twenty-two members representing a variety of local agencies and institutions, including business, government, the school systems, and others. Additionally, the project has enlisted the assistance of a sustainability consulting firm, located in Portland, Oregon, in designing GIS-based systems for assessing and monitoring development and its impacts.

San Jose, a California city of nearly 900,000 residents, has developed a fairly strong sustainability initiative. San Jose began the process of contemplating its approach to sustainability years before many other cities, as reflected in the 1980 report entitled "Toward a Sustainable City" prepared by city agencies for the city council (City of San Jose 1998b). By 1986, the city had created an Office of Environmental Management, located in the city manager's office. In 1993, the city council approved the creation of the Department of Environmental Services, which took over the functions of the OEM and functions from many other agencies. In August of 1994, San Jose's City Council adopted "San Jose 2020" as its general plan (City of San Jose 1998a). This plan included a provision entitled the "Sustainable City Major Strategy." The Sustainable City Major Strategy represented a statement of San Jose's desire to become an environmentally and economically sustainable city. Much like the programs profiled earlier in Seattle, Portland, and Boulder, sustainability is seen as a function of comprehensive planning. As a senior planner in the San Jose planning department stated, ". . . sustainability is built into our long-range planning. We are strong believers in long-range planning, and thinking of sustainability in this way allows us to build a strong policy foundation that will be critical to the success of our sustainability projects" (San Jose Planning Department 1999). In operation, implementation of San Jose's sustainability initiative falls to the responsibility of the city's Department of Environmental Services. It consists of efforts to integrate programs on solid waste recycling, air and water quality management, land use and growth management, transportation planning, and energy and water conservation. It also includes an indicators project, and the development of a green building program.

Some Correlates of Serious Sustainability Efforts

If the case descriptions begin to clarify what the concept of sustainability means in local U.S. contexts, they cannot do much to build a picture of why some cities seem to take the idea so much more seriously than others. In chapter 2, an effort was made to develop a single index of how seriously cities take sustainability, and to use this index to compare the twenty-four cities that have embarked on some form of sustainability initiative. Given the variations in local contexts that are evident in the case descriptions, it is clear that the meaning of the index must be circumscribed. Nevertheless, to the extent that the index represents a reasonable assessment of how seriously cities take sustainability, the index values can be treated as the dependent variable in an analysis to try to identify some basic correlates (Esty 2001). This analysis focuses expressly on the issue of why, among cities in the United States that have elected to pursue sustainability, do some seem to take the idea more seriously than others. This is a different question from the one that asks why, among all cities, do some embark on sustainability initiatives and others not. Answering this latter question would require a different sample of cities, one that perhaps compares the twenty-four cities examined here to some collection of cities without such initiatives. No such comparison is attempted here. The focus here is on a comparison of the twenty-four cities, all of which have already initiated some type of sustainability effort.

In chapter 7, a preliminary inquiry was started, looking into the correlates of sustainability, with particular focus on eight cities that seemed to be fairly serious. Here, the analysis is taken a major step further, presenting analysis based on the "Taking Sustainability Seriously Index" developed in chapter 2. Although the existing literature on sustainable cities or sustainable communities provides little in the way prediction or specific foundations on which hypotheses could be based, there are nonetheless many reasons why logically, or by extension from other explanations, one might expect some cities to be more serious about sustainability than others.

If it is possible to imagine that cities can take sustainability seriously, and if it is also possible that cities can vary in the extent to which they do so, it is also possible to contemplate why some cities are more serious about pursuing sustainability than others. In general, the literature on sustainable communities and sustainable cities provides virtually nothing in the way of conceptual guidance on what might explain this variance. The following analysis develops some basic information about possible correlates of the effort cities put into their sustainability initiatives. The rationale for some of these correlates comes mainly from logic and intuition, and for others comes from extension of existing conceptual work.

Intuition, perhaps, tells us that the scale of a city's environmental challenges might play a role, but the opposite may hold true as well. While very large cities might have a greater need to pursue sustainability, the scale of that need might constitute an immense obstacle as well. Smaller cities might be able to avoid the impediments posed by the large scale that our biggest cities face, but they also tend to lack the critical mass of interests to push issues of sustainability onto the public agenda. The variables that come the closest to tapping these issues are population size, population growth, land area (in square miles), and gross population density (residents per square mile). Among the twenty-four cities with sustainability initiatives, the 2000 population size ranges from a little over 42,000 in Olympia to over 1.3 million in Phoenix, with an average population size of 413,000 people, and without a doubt, the kinds of environmental issues faced by Phoenix are very different, by virtue of its size, than those confronted in Olympia. This leads to a hypothesis that among the twenty-four cities that have sustainability initiatives, those that are larger, that have experienced the greatest population growth, that have the smallest land area, and that have the highest population density will be the cities that take sustainability most seriously.

Table 8.1 shows the relationship between population size, along with other demographic variables, and the Index of Taking Sustainability Seriously. While the direction of the relationships is in the predicted direction, none of these correlations approaches statistical significance. Examination of the relationship between population size and the index, through perusal of scatterplots and additional correlational analysis, reveals that there is no curvilinear relationship either.

Another plausible explanation for why some cities take sustainability more seriously focuses on the nature of the local resources and needs of the city. Although the empirical literature on public support for environmental programs reports very little in the way of a relationship between personal income and such support, there remains the possibility that cities with greater resources will be better able to afford the "luxury" of pursuing sustainable development. Cities with serious social and economic problems, so the argument goes, would be less apt to place a high priority on sustainability, instead worrying about other high priority issues. For example, a city facing high unemployment would be more likely to opt for any kind of economic development they can get, while cities with lower unemployment may find it easier to be choosy about the kind of economic development they will allow. This leads to a hypothesis that among the twenty-four cities with sustainability initiatives, those with higher median family incomes, higher median house

Table 8.1
Correlations between the Index of Taking Sustainability Seriously and demographic characteristics in twenty-four cities

Independent variable	Pearson Correlation Coefficient	Significance
Total population, 2000	.136	.53
Total population, 1990	.111	.60
Total population, 1980	.067	.76
Population change %, 1980 to 1990	.165	.49
Population change %, 1980 to 2000	.141	.51
Total land area (square miles)	−.014	.95
Population density (Population per square mile)	.127	.55

values, lower rates of poverty, lower unemployment rates, and higher per capita governmental spending will be the cities that take sustainability most seriously.

Table 8.2 shows the correlations with these variables. The only statistically significant correlation is that between the index and the 1990 poverty rate, suggesting that among the twenty-four cities, those with greater poverty have had the most difficulty taking sustainability seriously. The cities with higher median family incomes, lower unemployment rates, and higher median house values tend to be the cities that take sustainability more seriously, although the coefficients are not statistically significant. The government spending variables, total and per capital local government expenditures, show little correlations with the index. There does seem to be a tendency for cities with greater resources and fewer economic problems, such as a low unemployment rate, to take sustainability more seriously.

These patterns suggest that the resource character of the local population does not play a particularly strong role in determining how seriously the city will take the pursuit of sustainability. But what about other local characteristics of the populations of the cities? Does the racial or ethnic character seem to play a role? Is there reason to think that the age structure of the city's population influences the pursuit of sustainability? If cities adhere to the logic of the Brundtland Commission's concern for future generations, then cities with larger populations of children or

Table 8.2
Correlations between the Index of Taking Sustainability Seriously and local resource characteristics in twenty-four cities

Independent variable	Pearson correlation coefficient	Significance
Median family income, 1990	.313	.14
Poverty rate, 1990	−.397	.05
Average unemployment rate, 1994–96–99	−.273	.20
Median house value, 1990	.302	.15
Total city government spending, 1990	.079	.71
Per capita government spending, 1990	−.127	.56

lower average age should pursue sustainability more aggressively than those with older populations. What about the employment structure in the city? Does the presence of a significant manufacturing base mean that the city will experience a greater need to be environmentally sensitive by virtue of the fact that manufacturing industries often place greater burdens on the environment? Or does reliance on manufacturing as a base of employment mean that the city will be hesitant to go too far for fear that pursuing sustainability might threaten the very source of economic well-being? The education of the local populace may also play a role, where we might expect a more educated population to be more attuned to the desirability of pursuing sustainability. This leads to the hypothesis that among the twenty-four cities with sustainability initiatives, those with higher employment in manufacturing and lower employment in service industries, with high African American and Hispanic populations, less well-educated populations, and larger contingents of older people will be the cities that take sustainability less seriously.

Table 8.3 presents the correlations for these variables. Here we see that cities with large African American and Hispanic populations are indeed

Table 8.3
Correlations between the Index of Taking Sustainability Seriously and population and employment characteristics in twenty-four cities

Independent variable	Pearson correlation coefficient	Significance
Percent African American, 2000	−.391	.06
Percent African American, 1990	−.340	.10
Percent Hispanic, 2000	−.278	.18
Percent Hispanic, 1990	−.241	.26
Percent under 18 years old, 1990	−.495	.01
Percent over 65 years old, 1990	.147	.50
Median age of the population, 1990	.621	.00
Percent high school graduate, 1990	.501	.01
Percent employed in manufacturing, 1990	−.547	.01
Percent employed in service sector, 1990	.094	.66

the cities that take sustainability less seriously. Whether it is because these cities face what they consider to be more pressing problems, or because members of these populations simply do not place a high value on trying to achieve sustainability is impossible to say here. The age of the populations of cities also seems to make a difference, but not necessarily in the predicted direction. Cities with larger proportions of their populations under the age of 18 are much less likely to take sustainability seriously, and cities with higher median aged populations are far more likely to take sustainability seriously. Education does work in the predicted direction; cities with better-educated populations tend to take sustainability more seriously. Finally, cities that are reliant on manufacturing for the employment of their residents are far less likely to vigorously pursue sustainability than cities less reliant on manufacturing. If cities that have a strong presence of manufacturing industries are the most in need of sustainability initiatives by virtue of the environmental threats that are often associated with them, then they are decidedly not the cities that take sustainability most seriously.

Is there a pattern where cities that are predisposed to look favorably upon the environment and environmental protection are more likely to pursue sustainability? There are no direct measures of such predispositions, but three proxy measures might help to provide a little insight into this type of issue. For example, some cities are very much oriented toward the use of public transportation, and others are heavily reliant on the automobile, particularly for commuting to work. We might expect those cities that have traditionally been more reliant on the automobile (cities with large proportions of commuters who drive to work alone) to be the cities that are less interested in pursuing sustainability, and those that have been more oriented toward public transit to take sustainability more seriously. Additionally, cities with local governments that tended to spend greater amounts of money environmentally related activities (water quality, parks and recreation, sanitation, etc.) should be cities that have opted to take sustainability more seriously. Another proxy measure of such predispositions might be how the local electorate votes. Although this is but a crude and incomplete indicator of how "progressive" the cities' populations are, one might equate voting for Democratic candidates as a proxy measure. For example, in the last two presidential elections, one might expect cities that tended to vote more for the

Democratic candidates, Clinton in 1996 and Gore in 2000, would be more concerned about environmental issues, and more attuned to sustainability.

One of the first reactions people seem to have when they see a list of the twenty-four cities is to take particular note of the cities that are located on the west coast. Indeed, the list contains a healthy dose of such cities, from Seattle and Portland in the Pacific Northwest, to San Francisco, San Jose, Santa Barbara, and Santa Monica further down the coast. If these cities are disproportionately likely to take sustainability seriously, and there is plenty of reason to think that they might (DeLeon 1992), then this would suggest that sustainable cities is largely a west coast phenomenon, raising the question of whether the concept is feasible elsewhere in the United States. The case descriptions demonstrated that indeed there are cities away from the west coast that seem to take sustainability seriously. But is there a way to examine the extent to which these exceptions represent a few aberrations? If the serious pursuit of sustainability is largely a west coast phenomenon with a few exceptions, then the cities' presence on the west coast should be strongly correlated with the Index.

This leads to the hypothesis that among the twenty-four cities with sustainability initiatives, those with higher proportions of their populations using public transportation to get to work, those with fewer people who drive to work alone, those that traditionally have spent more money (per capita) on their environments, those that tend to vote disproportionately for Democratic presidential candidates, and cities that are located on the west coast will be the cities that take sustainability more seriously.

These correlations are presented in table 8.4, and they show that there is little tendency for these "predisposition" variables to be related to the index. The only variable with a statistically significant correlation is the location on the west coast (which is just barely significant). Perhaps just as interesting is the fact that the correlation between the amounts of money the city government spends on sustainability (crudely measured by the sum of spending on health, sewerage, water quality, and parks and recreation) is positively (but not significantly) correlated with the index, but the correlations between per capita spending on the environment and the index is negative (but insignificant).

Table 8.4

Correlations between the Index of Taking Sustainability Seriously and measures of environmental predisposition in twenty-four cities

Independent variable	Pearson correlation coefficient	Significance
Percent drivers driving alone to work, 1990	.015	.74
Percent of commuters using public transportation, 1990	.075	.73
Total government spending on environment, 1997	.202	.34
Per capita spending on environment, 1997	−.171	.44
Location on the West Coast (CA, OR, WA)	.406	.05
Average percent democratic presidential vote, 1996–2000 +	.205	.36

Note: + Based on data for 22 of 24 cities.

With all of these bivariate correlations, it is somewhat difficult to get a sense of which of these variables might produce the strongest influences on the Index, particularly controlling for other influences. A simple OLS multiple regression analysis, including five of the independent variables that are more highly correlated with the Index, as presented in table 8.5, shows significant correlations with only two variables. The median age of the city's population and the percentage of the labor force employed in manufacturing industries persist as significant correlates, with older populations and fewer manufacturing employees being associated with cities that take sustainability more seriously. Controlling for these variables, the size of the African American population, the level of education, and location on the west coast appear to not be significantly correlated with the index. In other words, it is not being located on the west coast, per se, that seems to contribute to these cities taking sustainability more seriously. It is the age and education levels of those places that seem to matter more.

Discussion and Summary of the Empirical Analysis

Surprisingly few of the independent variables developed here were correlated with cities' level of seriousness of sustainability. Many of the

Table 8.5
OLS regression results showing correlates of the Taking Sustainability Seriously Index

	β	SE β	Beta	T	Significance
Percent African American, 2000	−.166	.084	−.348	−1.104	.065
Median age, 1990	1.257	.414	.547	3.034	.007
Percent employed in manufacturing, 1990	−.0048	.002	−.458	−3.148	.006
Percent high school graduates, 1990	.0455	.122	.067	.374	.713
Location on the West Coast	−3.164	2.86	−.207	−1.104	.285
Constant	−21.080	12.46	—	−1.692	.109

Notes:
Multiple R   .829
R Square   .687
Adjusted R Square   .595

Analysis of Variance

	DF	Sum of Squares	Mean Square
Regression	5	785.08	157.02
Residual	17	357.35	21.02

F = 7.47
Significance of F = .0007.

variables that would seem to offer the most plausible explanations were but weakly correlated with the number of sustainability programs and activities in the city. Population size and rapid growth in population, variables often believed to put pressure on local governments to manage, limit, or stop future growth, were not related to the seriousness of sustainability. The resource base of cities, measured by median family income, median house value, total and per capita local government spending, and the unemployment rate, were not related to the seriousness of the sustainability effort. The poverty rate did reveal a modest bivariate relationship, but in multivariate analysis, never emerged as significant. The general lack of a pattern suggests that there really is not a single kind of city that finds sustainability particularly appealing. It is not just liberal cities that take sustainability seriously, although there is a healthy dose of such cities among the twenty-four studied here. It is not just west coast cities that are serious about sustainability, although there is a healthy dose of those among the twenty-four as well. It isn't just affluent, predominantly White, well-educated cities that have the greatest interest in local sustainability.

The variables that did emerge as significant correlates were demographic characteristics—median population age, the percentage of the population below 18 years of age, the percentage high school graduates, and the percentage of African Americans residing in the city—and one indicator of the local economy—the percentage of the labor force employed in manufacturing industries. What this probably means is that the cities that need sustainability the most—cities that are reliant on relatively more polluting manufacturing industries as the base of employment, and cities with younger populations—are the cities that tend to take sustainability less seriously. If efforts to achieve sustainability through cities in the United States are to succeed, then greater attention will need to be paid to defining the conditions under which the most needy cities can take the idea seriously. To some degree, this may be a problem that will resolve itself over time, at least in the sense that cities may find it increasingly possible to develop serious sustainability initiatives as their manufacturing base diminishes, as it has in most U.S. cities.

The analysis presented here is somewhat limited by virtue of the small number of cases available for analysis. With just twenty-four cases, the scope of the empirical analysis is, by its very nature, rather limited. But the fact is that, with the caveats discussed earlier, this number essentially represents the population of *cities* pursuing sustainability in the country. Obviously, as more cities begin to establish sustainability initiatives, this problem will diminish and more extensive comparative analysis can be conducted. In the short term, the population of cases can only be expanded by following one of three research strategies. First, the analysis could consider venturing outside of the United States, particularly to Canada, where numerous cities such as Toronto, Winnipeg, Hamilton, and Vancouver, among others, have established sustainability initiatives, and to Western Europe, where the idea of sustainable cities really got its start. Future research will undoubtedly focus on systematic comparison of the U.S. and non-U.S. cities' sustainability initiatives.

Second, smaller places, such as Ithaca, New York; Stuart, Florida; and Granstville, Utah, could be included in the analysis. But clearly these smaller municipalities face distinctly different sustainability challenges than the much larger cities examined here. Adding such smaller places would probably take on overtones of comparing apples with oranges rather than apples with apples, which would likely limit the inferences that could be made as a result. Third, the definition of what constitutes a sustainability initiative could be reconsidered so that cities omitted from this analysis, but which still have some sort of environmental initiatives, could be included. This would mean that other cities around the United States could be scrutinized for the existence of programs and activities that are consistent with the pursuit of sustainability, and cities that have such programs could be included even if they do not operate these programs as part of a sustainability effort. Additionally, future analysis will expand its scope to compare these twenty-four early adopter cities to others around the country to determine why some cities pursue sustainability at all, while most cities in the United States do not.

A Research Agenda for Future Analysis

From the outset, this analysis focused on highlighting a number of specific questions that arise out of cities' sustainability initiatives, and that arise out of their relationship to the conceptual foundations of sustainability. These questions pose significant research challenges, and yet answers to them hold great promise for providing guidance to cities that may seek to develop sustainability initiatives. Given the descriptions of cities' sustainability efforts, it is clear that there are a large number of unanswered questions about how these programs got started, about how they maintain or fail to maintain themselves, about what they have been able to achieve programmatically, and about how effectively they can accomplish environmental and related improvements. A review and discussion of these research questions promises to provide a foundation for future research on sustainable cities.

Perhaps what comes through most clearly from the profiles is the need for even more systematic comparison of the cities that have sustainability initiatives. Efforts to compare cities are often difficult and fraught with methodological minefields. Yet the case profiles and descriptions, and the preliminary correlational analyses presented earlier in this chapter, raise a variety of research questions that can only be addressed through rigorous in-depth and comparative city analysis. Are there other common characteristics among the cities that have such programs? In what ways are the cities that have started sustainability programs different from cities that have not? Are the observable differences among the sustainable cities initiatives simply a function of different local political styles and contexts, or are there substantive differences in terms of how likely they are to produce measurable environmental impacts? If the local governance regime in the city matters, are there some general rules concerning what kinds of regimes are most supportive of sustainability? Perhaps more important, are there ways that such regimes can be built or created in cities where they do not now exist? These and related issues are elaborated next. Since systematic comparison of cities is facilitated by some sort of standard measure or set of criteria of the seriousness of cities' sustainability efforts, that is where the discussion begins. In chapter 2, the outline of the measurement used in this book was presented, and the criteria were made explicit. Can further progress be made to do a better job of measuring how serious cities' sustainability efforts are? Are there elements of sustainability whose measurement needs to be refined or altered? This forms a starting point for defining a future research agenda.

The Measurement of "Taking Sustainability Seriously"

One lesson that emerges from this analysis is the fact that taking sustainability seriously is a fairly complex concept, and must be viewed as multidimensional. Obviously the environment plays a significant, if not the most important, role in any such analysis. But as the preceding discussions demonstrated, sustainability also relates to energy consumption, transportation and land use planning, community building, and perhaps environmental and social justice issues. Equally important is the obvious conclusion that it is very difficult to know what it means to take sustainability seriously, particularly in a way that separates this concept from achieving actual, tangible, results. Perhaps at future times it will make sense to incorporate measures of such results into measures of how seriously cities take sustainability. For the purposes of this book, the effort was mainly focused on understanding the extent to which the city was trying to become more sustainable, particularly with respect to the city's government, its public policies and programs, and its various non-profit organizations. Yet because of the complexity of cities' programs, and because the programs look so different from city to city, judgments about whether one city seems to take sustainability more seriously than others, or judgments about how much more seriously one city seems to take sustainability than others, are extremely difficult to make.

As implied earlier, there may come a day when research will provide greater clarity on the relationship between the programs and actions of cities and the environmental quality consequences of these programs. At that point, a city might be said to take sustainability seriously when its program is found to produce environmental improvements, and one city's initiative might be said to be more serious than other cities' initiatives if it produces greater environmental benefit. Until this kind of research is feasible, however, the measurement of taking sustainability seriously will inevitably be based more on judgment than on rigorous objective standards. This raises an additional question, whether the measurement of taking sustainability serious is based on subjective or objective standards, concerning whether it is possible to develop a single index as an indicator.

Research on the Causes and Correlates of "Taking Sustainability Seriously"

This analysis sought to begin the process of examining, in a preliminary way, some reasons why cities would engage in efforts to become more sustainable, and for reasons why some cities would take sustainability more seriously than others. Clearly, this search needs to continue. There are few clear-cut patterns of association with characteristics of the cities—indeed, the cities that seem to be taking sustainability seriously differ greatly from one another. Some are large, and some are relatively small. Some experienced very rapid population growth, and some much less so. Some faced significant environmental problems, and some have not. Some are very densely populated areas, and some are less so. So are on the west coast, but many are not. Some are politically liberal places, and some are pretty conservative. In short, the broad-brush characteristics do not seem to characterize very well the cities that take sustainability seriously.

So the next logical question focuses on whether there are other characteristics that the sustainable cities share in common and that make them different from other places that ostensibly take sustainability less seriously. Are there specific characteristics of the local economies that help explain cities' decisions to engage in sustainable development? It is certainly possible that various kinds of economic displacements have made smart growth look more inviting to cities. It might be hypothesized that cities that have lost disproportionate amounts of the employment base, particularly through the loss of manufacturing industries, and cities whose personal incomes have declined in real terms would be more likely to aggressively look for fresh alternatives. As cities confront new challenges to traditional economic development, and the unanticipated consequences of these challenges, they may well begin to look for alternative approaches to economic development. Smart growth, particularly in the context of a sustainability initiative, provides such alternatives and approaches.

Yet the opposite hypothesis is also highly plausible. In other words, it could be that cities without significant local economic displacement, and with adequate local resource bases, are those that have been able to experiment with sustainability. Although there is no general pattern where personal income, at the individual level, correlates with stronger support for protecting and improving the environment, there is certainly the possibility that cities with greater resources will feel less constrained to engage in efforts to protect their local environments. Indeed, the relationship between local economic performance and likelihood of taking sustainability seriously need not be a linear one. It could be that both kinds of cities—cities that have had rough economic times and cities that have been relative well off—would both find advantages to pursuing sustainability initiatives.

There is little question that the character of local politics plays a role in determining the organization and form the sustainability initiative takes. Certainly, the Southern cities studied here were much more likely to rely on nonprofit organizations to serve integral parts in making their sustainability initiatives happen. Many of the specific program elements in Chattanooga have been put in place by nonprofit organizations, and the Chamber of Commerce has been a very active player. The political conditions under which a city might prefer to rely on local government versus nonprofit organizations need to be delineated. Does the dominant political ideology of the city in some way determine which "model" cities decide to rely on? Moreover, if Brugmann is correct in his assertion that serious sustainability initiatives need to be developed and managed by city governments to be effective, then this raises the question of whether politically conservative cities can ever take sustainability seriously. The analysis here certainly suggests that there are limits to how far, and in what ways, relatively politically conservative cities can go in defining a sustainability initiative. But cities such as Chattanooga, Jacksonville, Austin, and others have perhaps done considerably more than their dominant political ideologies might predict. However, questions concerning the political preconditions for engaging in serious sustainability efforts remain unanswered.

Additionally, the preceding analysis raised questions about the relationship between local political participation and sustainable cities initiatives. It is clear that many cities' indicators projects were developed with an eye toward engaging residents in the process, and of measuring the extent of participation on an ongoing basis. Often, the indicators projects include emphasis on participation because of an unstated expectation that such participation will produce a more knowledgeable public, a public that is willing to be more supportive of cities' efforts to promote sustainability activities and programs. Stated in another way, the theory was that engaging the public would represent an important causal element in getting sustainability on the local political agenda and keeping it there. Yet there is surprisingly little empirical evidence to support the idea that this works. We might hypothesize that cities with more effective and extensive participatory processes will be able to generate the political support for enacting and implementing sustainability programs in their local governments. We might also hypothesize that after a sustainability initiative gets under way, cities with greater public participation in the initiative will be better able to sustain that initiative over time than cities with less participation. This idea is discussed more fully next.

The Role of the Business Community in Sustainable Cities

As described largely in chapter 4, the sustainability initiatives in many cities try to incorporate the business community and business-related issues into their activities. Sometimes this takes the form of city agencies providing technical assistance to private, usually small, businesses. This is the case with the green building technical assistance programs in Austin, Boulder, and many other cities. It also characterizes the activities of the Partners for a Clean Environment Program in Boulder. Sometimes it takes the form of creating nonprofit community development corporations. Sometimes it involves developing partnerships with specific business organizations, particularly local or area-wide chambers of commerce, as was found in Chattanooga and Boulder. Whatever form it might take, the business community promises to become an integral part of cities' sustainability activities.

What is less clear is whether one form of business involvement would seem to work better than others in promoting sustainability. Stated another way, it is not clear exactly what role the business community should play, must play, or probably would not be willing to play in defining the sustainability agenda. It may be difficult to imagine a city making great progress toward becoming more sustainable without the intimate involvement of local business and industry. Yet previous chapters demonstrated that there is a clear tension between the involvement of business and proponents of sustainability, whether proponents work within city government or in an independent organization. This tension is perhaps grounded in concern that involvement of business and industry will inevitably undermine or co-opt the sustainability initiative. Most of the sustainable indicators projects do not explicitly seek to engage the business community, although there certainly are some notable exceptions. Part of the reason why the business community tends to be largely excluded would appear to be the worry that the entire endeavor would be taken over to benefit that community perhaps at the expense of progress toward sustainability. In Chattanooga, the possibility exists that the elevated role of the Chamber of Commerce in consolidating many of the community development activities previously conducted by other organizations could represent an effort to co-opt sustainability purely for the purpose of advancing economic development in the form of cluster development. Since it is not clear what, if anything, cluster development promises to do for sustainability, the implication is that it may do nothing.

On the other hand, it is clear that the more well-defined the role of the business community is, the less likely the sustainability program is to take aggressive measures to define or implement sustainability initiatives that might negatively affect specific business interests. It is difficult for cities that seek to aggressively apply land use controls for the purpose of achieving smart growth to stay engaged with the real estate development community. To the business world, less regulation, including less regulation of land use, is better. When cities sustainability initiatives seek to regulate business, that is often too much for business interests to swallow. The issue of the relationship between the business community and sustainability initiatives is really part of a much larger set of issues concerning the character of the urban regime. Stated another way, the question is whether, consistent with Clarence Stone's formulation (1993), some urban regime types are better able than others to accommodate and promote sustainability without undermining its fundamental purpose. Moreover, there is a great need for empirically based prescriptions to suggest what kinds of processes might be said to be more effective in transforming local governance regimes so that they are better able to accommodate the pursuit of sustainability.

Issues of the role of the business community raise a variety of specific hypotheses that need to be examined. One might examine, for example, the hypothesis that cities whose sustainability initiatives keep an arm's-length relationship with business and industry will be able to take sustainability more seriously than cities where the business community forms the central core of the sustainability initiatives. Of course, a counterargument can also be put forth that the business community must be an integral part of sustainability, leading to an opposite hypothesis that cities whose initiatives thoroughly incorporate the business community and its concerns will be more likely to take sustainability seriously than those that do not. Alternatively, some limited role for the business community may lead to the hypothesis that cities whose initiatives provide for limited participation of the business community will be more likely to take sustainability seriously than cities whose initiatives either ignore the business community or cities where the business community forms the central core of the initiative.

These issues concerning the role of the business community raise addition questions concerning the composition and operation of urban regimes that would appear to be most capable of providing the political support necessary for cities to pursue sustainability. Although environmentalists and business interests would not normally be thought of as natural allies in local policymaking, there are enough examples from the cases examined here to suggest that it is possible for such alliances to be formed. This analysis has intimated at times that such alliances are possible, but the descriptions are somewhat short on details. Any agenda for future research on sustainable cities must take a much harder look at how such alliances are formed, what their bases and common interests are, and above all, the extent to which any particular vision of sustainability might be sacrificed. Under what conditions do the business communities of cities become supportive of sustainability initiatives, and what, if anything, is lost in order to secure that support?

Research on the Consequences of "Taking Sustainability Seriously"

Although analysis of what causes some cities to take sustainability more seriously than others represents an important research question, it is also important to begin to evolve an understanding of the relationship between the sustainability programs and policies of cities and the quality of the environment or the quality of life. This book began with the notion that it is premature to examine this question simply because even the most optimistic visions of sustainability anticipate that it might take many years for cities to reap the benefits of their programs. Yet at some point, it is necessary to build a base of knowledge about what cities are getting for the effort they put into their sustainability initiatives.

Perhaps the most important reason for developing such a base of knowledge is to address the overriding issue concerning whether cities can do anything to affect sustainability. As discussed in chapter 1, there are many people who are skeptical that any city, acting alone, can make even the smallest dent in the environmental health of the world. Of course, optimists believe that a city can be part of a collective urban effort to make a difference. In the future, it will become increasingly important to ascertain whether the optimists are correct. Although the temptation may be to simply compare cities with sustainability initiatives to cities without such initiatives over time, usable information is more likely to result from time series assessments within cities, particularly assessments that focus on the effects of specific programs or policies. Questions such as "did Boulder's voluntary VOC reduction program produce measurable declines in VOC emissions from 2000 to 2010," or "did Austin's elective renewable energy source option reduction reliance on fossil fuels from 2000 to 2005," or other program-specific questions, promise to build a detailed understanding of what kinds of programs seem to work. Ultimately, it makes no sense for a city to try to emulate the program of another city if the program does not demonstrably produce the desired results.

One of the promises offered by the creation of a single index that measures the extent to which cities take sustainability serious, as discussed earlier, is that it might facilitate analysis of the relationship between cities programs, policies and actions on one hand, and actual improvements in the environment. Whether analysis tries to rely on such an index or not, there is probably no issue of greater importance than the relationship between cities' sustainability initiatives and the creation of healthier, more livable, and more sustainable places. No one would bother to develop sustainable indicators if there was not an expectation that doing so will contribute to environmental improvements. No one would advocate and establish green building programs if they did not believe such programs actually improve the environment. Few people would support smart growth if the end result were to not help protect the environment. In short, sustainability initiatives promote a wide range of city activities, programs, and policies because of the expectation that these programs will pay off.

As matter for research, this suggests a number of hypotheses. One might expect, in general, that cities that take sustainability seriously will be more likely to experience improvements in the environment than cities that do not take sustainability seriously. Alternatively, cities that take sustainability more seriously will experience greater improvements (or less degradation) in the environment than cities that take sustainability less seriously. Additional specific hypotheses might suggest that cities with specific program elements or types would experience improvements in specific types of environmental quality. For example, one might expect that cities with green building programs to use less energy and less water per capita than cities without such programs; cities with VOC reduction programs will have fewer air quality problems than cities without VOC reduction programs. The specific program elements also suggest evaluative hypotheses that apply to specific cities over time. For example, one might expect that after adopting a VOC reduction program, a city should experience fewer air quality problems than before the program was adopted.

Finally, there continue to be questions concerning the effects of city-based growth limit strategies. These questions focus on whether a city's efforts to limit growth produce the desired effect within the city or within a metropolitan area, whether they produce unintended consequences, or whether they turn out to be counterproductive. There are certainly studies that examine the hypothesis that cities that adopt growth limit policies are less likely to experience economic growth than cities that do not. However, such studies do not directly address the issues that such policies are designed to target. Few people would argue that growth limit policies are benign with respect to the environment. The question is whether such policies accomplish their environmental and livability goals. Here the research is much less evident. Tests of the hypothesis that cities using growth limits are more likely to experience improved environmental quality (or less degradation of the environment) than cities without growth limits have rarely been conducted.

At least one observer has proposed the hypothesis that growth limits will actually undermine the pursuit of sustainability in a city. Anthony Downs (2000) suggests in a November 2000 policy brief that without understanding population growth limitation within the context of the region in which a city is located could make problems worse. As he states it, "... when one locality passes laws limiting future growth within its own boundaries, it does not affect the future growth rate of the overall region, but rather moves the region's future growth to other localities, or to outlying unincorporated areas. Local growth limits, then, actually aggravate sprawl" (p. 4).

Logically, this seems like a plausible consequence of growth limitation measures. But there has been little effort to examine this as an empirically testable hypothesis. If Downs is correct, one might expect that metropolitan areas with cities that have adopted growth limitation measures but with no comparable growth limitation measures in the remainder of the areas, will experience greater development sprawl than metropolitan areas that have no growth limitation measures. The fact is, however, that most of the cities examined here that engage in growth management through comprehensive planning do so in the context of state-wide planning initiatives. Comprehensive growth management in Seattle and Olympia do not take place in the kind of vacuum described by Downs, but rather in the context of metropolitan, regional, and state-wide planning. The same is true of the sustainable cities in Arizona and Florida. From the perspective of sustainability, however, the point is that there really is no empirical analysis that examines the relationship between the effectiveness of cities' growth management efforts and the efforts of their surrounding metropolitan areas or substate regions.

The Role of Indicators Projects in Building Social and Political Capital

If one of the justifications for the development of a sustainable indicators project is that it will contribute to the creation of the social and political capital in support of sustainability, there is surprisingly little evidence that this has happened. As the preceding analysis suggests, there are specific cases where one can see the possibility that the processes used in association with an indicators project were successful in that effort. The efforts of Sustainable Seattle are perhaps a case in point. Threaded throughout advocacy of sustainable cities' initiatives is the idea that engaging people in the city in the process of indicator development pays political dividends. Whether through public hearings, or "charettes", or through some other "visioning" process, the implicit expectation is that the political support for pursuing sustainability programs through city government will emerge from participation. Yet there is no systematic evidence that indicators projects in general, or even highly participatory indicators projects, contribute to the fabric of the social and political environment of sustainability. Moreover, there is even less information concerning what kinds of participation, and how much participation, might be said to maximally contribute to making the city safe for sustainability.

Additionally, there is an expectation that participatory processes associated with indicators projects will facilitate placing sustainability issues on the public agenda. The idea is that once the indicators project is successful in engaging residents of a city, the mainstream political leaders of the city will find it difficult to ignore sustainability or to pursue policies and programs that operate at counterpurposes to sustainability. Again, there is very little empirical evidence on this posited relationship. As one hypothesis, one might suggest that cities with sustainable indicators projects will be more likely to incorporate sustainability issues into the public agenda than cities without indicators projects. In other words, government agencies in cities with indicators projects are more likely to adopt programs and pursue activities that are consistent with sustainable development than agencies in cities without indicators projects.

Another implicit expectation often embedded in sustainability initiatives concerns the contribution of any broad-based participatory processes associated with sustainability initiatives to a changed system of governance. The expectation, as outlined in chapter 5, is that these participatory processes will help to overcome the three deadly sins—the three systemic challenges to sustainability. So the question becomes can the participatory processes associated with sustainability initiatives help to overcome the tragedy of the commons, the NIMBY syndrome as it applies to environmentally impacting activities, and transboundary shifting of environmental impacts? Although communitarian theorists offer plausible rationales for how this could happen, there is surprisingly little direct empirical evidence that participatory processes conducted in the context of sustainable cities initiatives actually accomplish any of these goals. The problem, from an empirical perspective, is that this aspect of sustainable cities has received very little attention in research. The hypotheses that cities with participatory sustainability initiatives will solve common pool resource issues with less adversarial conflict, will avoid the paralysis associated with NIMBY opposition to siting important facilities that promise to improve the environment, and will confront (and avoid) shifting their environmental impacts outside of their cities, deserve to be examined directly.

References

Chapter 1

Bank of America. 1995. *Beyond Sprawl: New Patterns of Growth to Fit the New California*. San Francisco, Calif.: Bank of America.

Barton, Hugh. 1998. Eco-Neighborhoods: A Review of Projects. *Local Environment* 3, no. 2 (June): 159–177.

Beatley, Timothy. 2000. *Green Urbanism: Learning from European Cities*. Washington, D.C.: Island Press.

Beatley, Timothy, and Kristy Manning. 1997. *The Ecology of Place: Planning for Environment, Economy, and Community*. Washington, D.C.: Island Press.

Bell, Colin, and Howard Newby. 1974. *The Sociology of Community: A Selection of Readings*. London: Frank Cass Publishers.

Berry, Jeffrey M. 1999. *The New Liberalism: The Rising Power of Citizen Groups*. Washington, D.C.: Brookings Institution Press.

Brown, Becky J., Mark E. Hanson, Diana M. Liverman, and Robert W. Meredith. 1987. Global Sustainability: Toward Definition. *Environmental Management* 11, no. 6: 713–719.

Bryner, Gary. 2000. *Gaia's Wager: Environmental Movements and the Challenge of Sustainability*. Summit Ridge, N.J.: Rowman and Littlefield.

Building Livable Communities: A Report from the Clinton Gore Administration. 1999. Washington, D.C.: Livable Communities Initiative, June.

Calthorpe, Peter, and William Fulton. 2001. *The Regional City*. Washington, D.C.: Island Press.

Carley, Michael, and Philippe Spapens. 1998. *Sharing the World: Sustainable Living and Global Equity in the 21st Century*. London: Earthscan Publications.

Cassidy, Robert. 1980. *Livable Cities*. New York: Holt, Rinehart, and Winston.

Diamond, Henry L., and Patrick E. Noonan. 1996. *Land Use in America*. Washington, D.C.: Island Press.

Dryzek, John. 1987. *Rational Ecology: Environment and Political Economy*. New York: Basil Blackwell.

Dunphy, Robert T., Deborah Brett, Sandra Rosenbloom, and Andre Bald. 1996. *Moving beyond Gridlock: Traffic and Development.* Washington, D.C.: Urban Land Institute.

Fainstein, S. S., and N. I. Fainstein. 1986. Regime Strategies, Communal Resistance, and Economic Forces. In *Restructuring the City: The Political Economy of Urban Redevelopment,* N. I. Fainstein, S. S. Fainstein, R. C. Hill, D. Judd, and M. P. Smith (eds.), pp. 245–288. New York: Longman.

Fox, Jonathan A., and David L. Brown, eds. 1998. *The Struggle for Accountability: The World Bank, NGOs, and Grassroots Movements.* Cambridge, Mass.: MIT Press.

Gilman, Diane, and Richard Gilman, eds. 1991. *Eco-Villages and Sustainable Communities.* Langley, Wash.: The Context Institute.

ICLEI. 2000. *Local Agenda 21 Model Communities Programme.* Toronto, ON: International Council for Local Environmental Initiatives. Found at: <http://www.iclei.org/la21/la21updt.htm>.

Kidd, Charles V. 1992. The Evolution of Sustainability. *Journal of Agricultural and Environmental Ethics* 5, no. 1: 1–26.

Lachman, Beth E. 1997. *Linking Sustainable Community Activities to Pollution Prevention: A Sourcebook.* Santa Monica, Calif.: Critical Technologies Institute, RAND.

Leo, Christopher. 1998. Regional Growth Management Regime: The Case of Portland, Oregon. *Journal of Urban Affairs* 20: 363–394.

Marvin, Simon, and Simon Guy. 1998. Creating Myths Rather than Sustainability: The Transition Fallacies of the New Localism. *Local Environment* 2, no. 3 (October): 311–318.

Mazmanian, Daniel, and Michael Kraft, eds. 1999. *Toward Sustainable Communities: Transition and Transformations in Environmental Policy.* Cambridge, Mass.: MIT Press.

Milbrath, Lester. 1984. *Environmentalists: Vanguard for a New Society.* Albany, N.Y.: SUNY Press.

Moe, Richard, and Carter Wilkie. 1997. *Changing Places: Rebuilding Community in the Age of Sprawl.* New York: Henry Holt and Company.

Molotch, Harvey. 1976. The City as Growth Machine: Toward a Political Economy of Place. *American Journal of Sociology* 82, no. 2: 309–332.

National Commission on the Environment (NCE). 1993. *Choosing a Sustainable Future.* Washington, D.C.: Island Press.

National Science and Technology Council (NSTC). 1995. *Bridge to a Sustainable Future: National Environmental Technology Strategy.* Washington, D.C.: NSTC.

Odum, E. P. 1983. *Basic Ecology.* New York: Saunders College Publishing.

Partners for Livable Communities. 2000. *The Livable City: Revitalizing Urban Communities.* New York: McGraw-Hill.

Pierce, J. T., and A. Dale. 1999. *Communities, Development, and Sustainability across Canada.* Vancouver, British Col.: University of British Columbia Press.

Pye-Smith, Charlie, and Grazia Borrini Feyerabend. 1994. *The Wealth of Communities: Stories of Success in Local Environmental Management.* West Hartford, Conn.: Kumarian Press.

Rees, William E. 1996. Revisiting Carrying Capacity: Area-Based Indicators of Sustainability. *Population and Environment* 17, no. 3: 195–215.

Rees, William E. 1992. Ecological Footprints and Appropriated Carrying Capacity: What Urban Economics Leaves Out. *Environment and Urbanization* 4, no. 2: 121–130.

Rees, William E., and Mathis Wackernagel. 1994. Ecological Footprints and Appropriated Carrying Capacity: Measuring the Natural Capital Requirements of the Human Economy. In *Investing in Natural Capital: The Ecological Economics Approach to Sustainability,* A-M. Jannson, M. Hammer, C. Folke, and R. Costanza (eds.), pp. 362–390. Washington, D.C.: Island Press.

Rees, William E. 1997. Is 'Sustainable City' an Oxymoron? *Local Environment* 2, no. 3: 303–310.

Robinson, John, George Francis, Russel Legge, and Sally Lerner. 1990. Defining a Sustainable Society: Values, Principles, and Definitions. *Alternatives* 17, no. 2: 36–46.

Satterthwaite, David. 1997, Sustainable Cities or Cities that Contribute to Sustainable Development? *Urban Studies* 34, no. 10: 1667–1691.

Selman, Paul. 1996. *Local Sustainability: Managing and Planning Ecologically Sound Places.* New York: St. Martin's Press.

Shaiko, Ronald G. 1999. *Voices and Echoes for the Environment: Public Interest Representation in the 1990s and Beyond.* New York: Columbia University Press.

Stone, Clarence N. 1989. *Regime Politics: Governing Atlanta 1946–1988.* Lawrence, Kans.: University Press of Kansas.

Stone, Clarence N. 1993. Urban Regimes and the Capacity to Govern. *Journal of Urban Affairs* 15, no. 1: 1–28.

United Nations Environmental Programme. 2000. *Agenda 21.* New York: United Nations. Found at: <http://www.unep.org/Documents/Default.asp?DocumentID=52>.

USEPA. 1999. *EPA's Framework for Community-Based Environmental Protection.* Washington, D.C.: U.S. Environmental Protection Agency, February.

Waste, Robert. 1989. *The Ecology of City Policymaking.* New York: Oxford University Press.

World Commission on Environment and Development (WCED) (Brundtland Commission). 1987. *Our Common Future.* New York: Oxford University Press.

Zachary, Jill. 1995. *Sustainable Communities Indicators: Guideposts for Local Planning.* Santa Barbara, Calif.: Community Environmental Council, Inc.

Chapter 2

AtKisson, Alan. 1996. Developing Indicators of Sustainable Community: Lessons from Sustainable Seattle. *Environmental Impact Assessment Review* 16: 337–350.

Beatley, Timothy, and Kristy Manning. 1997. *The Ecology of Place: Planning for Environment, Economy, and Community.* Washington, D.C.: Island Press.

Brugmann, Jeb. 1997a. Is There a Method in Our Measurement? The Use of Indicators in Local Sustainable Development Planning. *Local Environment* 2, no. 1: 59–72.

Brugmann, Jeb. 1997b. Sustainability Indicators Revisited: Getting From Political Objectives to Performance Outcomes—A Response to Graham Pinfield. *Local Environment* 2, no. 3: 299–302.

Central Texas Indicators 2000. 2000. *Sustainability Indicators Project of Hays, Travis, and Williamson Counties: A Report on the Economic, Environmental, and Social Health of the Central Texas Region.* Found at: <http://www.centex-indicators.org/report.html>.

City of Portland. 1999. *Comprehensive Plan Goals and Policies.* Portland, Ore.: Bureau of Planning. Found at: <http://www.planning.ci.portland.or.us/pdf/ComprehensivePlan.pdf>.

City of Santa Barbara. 1999. *Community Indicators.* Santa Barbara, Calif.: South Coast Community Indicators Project.

City of Scottsdale. 2000. *Scottsdale Seeks Sustainability: 2000 Indicators Report.* Scottsdale, Ariz.: Environmental Planning and Design Division. Available at: <http://www.ci.scottsdale.az.us/environmental/Sustainable/>.

City of Seattle. 2000. *Seattle Draft Comprehensive Plan.* Seattle, Wash.: Strategic Planning Office, Department of Comprehensive and Neighborhood Planning. Found at: <http://www.cityofseattle.net/planning/Comprehensive/HOMECP.htm>.

Corson, Walter. 1992. *Priorities for Sustainable Societies.* Washington, D.C.: Global Tomorrow Coalition, July. Reprint.

Corson, Walter. 1993, *Measuring Urban Sustainability.* Washington, D.C.: Global Tomorrow Coalition, November. Reprint.

Esty, Dan. 2001. *The Environmental Sustainability Index.* New York: Global Leaders for Tomorrow Environment Task Force, World Economic Forum. Found at: <http://www.ciesin.columbia.edu>.

Global Cities Online. 1998. *Indicators of Sustainability.* Found at: <http://www.globalcities.org/intropg/demo/PGI/indicat.html>.

Hall, Bob, and Mary Lee Kerr. 1991. *Green Index 1991–1992: A State by State Guide to the Nation's Environmental Health.* Washington, D.C.: Island Press.

ICMA. 2001. *Smart Growth Network.* Washington, D.C.: International City/County Management Association. Found at: <http://www.smartgrowth.org/>.

Jacksonville Community Council, Inc. 1999. *Quality of Life in Jacksonville: Indicators for Progress*. Found at: <http://www.jcci.org/qol/qol.htm>.

Kline, Elizabeth. 1995a. *Sustainable Community Indicators*. Medford, Mass.: Consortium for Regional Sustainability Tufts University, January. Reprint.

Kline, Elizabeth. 1995b. *Sustainable Community Indicators: Examples from Cambridge, MA*. Medford, Mass.: Consortium for Regional Sustainability Tufts University, February. Reprint.

Lachman, Beth E. 1997. *Linking Sustainable Community Activities to Pollution Prevention: A Sourcebook*. Santa Monica, Calif.: Critical Technologies Institute, RAND.

MacLaren, V. W. 1996. Urban Sustainability Reporting. *Journal of the American Planning Association* 62, no. 2: 184–202.

Newman, Peter, and Jeffrey Kenworthy. 1999. *Sustainability and Cities: Overcoming Automobile Dependence*. Washington, D.C.: Island Press.

Pinfield, Graham. 1997. The Use of Indicators in Local Sustainable Development Planning: A Response to Jeb Brugmann. *Local Environment* 2, no. 2: 185–187.

Putnam, Robert D. 2000. *Bowling Alone*. New York: Touchstone Press, Simon and Schuster.

Rees, William E. 1992. Ecological Footprints and Appropriated Carrying Capacity: What Urban Economics Leaves Out. *Environment and Urbanization* 4, no. 2: 120–130.

Rees, William E. 1996. Revisiting Carrying Capacity: Area-Based Indicators of Sustainability. *Population and Environment* 17, no. 3: 195–215.

Rees, William E., and Mathis Wackernagel. 1994. Ecological Footprints and Appropriated Carrying Capacity: Measuring the Natural Capital Requirements of the Human Economy. In *Investing in Natural Capital: The Ecological Economics Approach to Sustainability*, A-M. Jannson, M. Hammer, C. Folke, and R. Costanza (eds.), pp. 362–390. Washington, D.C.: Island Press.

Sustainable Chattanooga. 1995. *The PCSD Briefing Book*. Found at: <http://www.chattanooga.net/sustain/pcsd_briefing_book/outline.html>.

Sustainable Seattle. 1992–93. *Proposed Key Indicators of Sustainable Community*. Seattle, Wash.: Sustainable Seattle Indicators Project, Version 5, December 1992, and Version 6, January 1993. Reprint.

Swanstrom, Todd, and Dennis R. Judd. 1998. *City Politics: Private Power and Public Policy*. 2d edition. New York: Longman.

Wackernagel, Mathis. 1998. The Ecological Footprint of Santiago de Chile. *Local Environment* 3, no. 1: 7–25.

World Resources Institute. 1993. *The 1993 Information Please Environmental Almanac*. Boston, Mass.: Houghton Mifflin.

Zachary, Jill. 1995. *Sustainable Community Indicators: Guideposts for Local Planning*. Santa Barbara, Calif.: Community Environmental Council, Inc.

Chapter 3

Baltimore Brownfields Project. 2000. *Brownfields Pilot–Baltimore*. Washington, D.C.: U. S. Environmental Protection Agency, Office of Solid Waste and Emergency Response. Found at: <http://www.epa.gov/brownfields/pdf/sc_balti.pdf>.

City of Austin. 2000. *The Sustainable Building Source Book*. Austin, Texas: City of Austin. Also see "The Green Building Program," found at: <http://www.ci.austin.tx.us/greenbuilder/programs.htm>.

City of Boulder. 2000. *Green Points Program Regulations*. Boulder, Colo.: City of Boulder, Department of Building Services. Found at: <http://www.ci.boulder.co.us/buildingservices/codes/greenpoints/greenpoints.htm>.

City of San Francisco. 1995. Goals, Objectives, Actions, and Indicators. *Sustainable City: Working Toward a Sustainable Future for San Francisco*. San Francisco, Calif.: Sustainable City San Francisco. Found at: <http://www.sustainable-city.org/document/backgr2.htm>.

City of San Francisco. 1996a. Biodiversity. *Sustainability Plan for San Francisco*. San Francisco, Calif.: San Francisco Commission on the Environment. Found at: <http://www.sfgov.org/sfenvironment/environment/sustain/biodvrst.htm>.

City of San Francisco. 1996b. Energy, Climate Change, and Ozone Depletion. *Sustainability Plan for San Francisco*. San Francisco, Calif.: San Francisco Commission on the Environment. Found at: <http://www.sfgov.org/sfenvironment/environment/sustain/energy.htm>.

City of San Jose. 2000. *Green Building Program*. San Jose, Calif.: Environmental Services Department. Found at: <http://www.ci.san-jose.ca.us/esd/gb-home.htm>.

City of Seattle. 2000. *Seattle Draft Comprehensive Plan*. Seattle, Wash.: Strategic Planning Office, Department of Comprehensive and Neighborhood Planning. Found at: <http://www.cityofseattle.net/planning/comprehensive/HomeCP.htm>.

Dunphy, Robert T., Deborah Brett, Sandra Rosenbloom, and Andre Bald. 1996. *Moving beyond Gridlock: Traffic and Development*. Washington, D.C.: ULI-Urban Land Institute.

Greenlights Foundation. 1999. *Eco-Industrial Park Forum: A Review*. Newport, R.I.: Greenlights Foundation, Inc., November 3. Found at: <http://users.efortress.com/greenltfnd/forum.html>.

Hart, S., and A. L. Spivak. 1993. *The Elephant in the Bedroom: Automobile Dependence and Denial*. Pasadena, Calif.: New Paradigm Press.

Liroff, Richard A. 1976. *A National Policy for the Environment: NEPA and Its Aftermath*. Bloomington: Indiana University Press.

MacKenzie, W. R., N. J. Hoxie, M. E. Proctor, S. M. Gradus, K. A. Blair, D. E. Peterson, J. J. Kazmierczak, D. G., Addiss, K. R. Fox, J. B. Rose, and J. P. Davis. 1994. A Massive Outbreak in Milwaukee of Cryptosporidium Infection

Transmitted through the Public Water Supply. *New England Journal of Medicine* 331: 161–167.

Newman, Peter, and Jeffrey Kenworthy. 1999. *Sustainability and Cities: Overcoming Automobile Dependence.* Washington, D.C.: Island Press.

Ophuls, William, and A. Stephen Boyan, Jr. 1992. *Ecology and the Politics of Scarcity Revisited.* New York: W. H. Freeman.

PACE. 2000. *Partners for a Clean Environment.* Boulder, Colo.: Department of Environmental Affairs. Found at: <http://www.ci.boulder.co.us/environmentalaffairs/PACE/business_menu.html>.

Portland Brownfields Initiative. 2000. *Brownfields Showcase Community: Portland, OR.* Washington, D.C.: U. S. Environmental Protection Agency, Office of Solid Waste and Emergency Response. Found at: <http://www.epa.gov/brownfields/pdf/sc_portl.pdf>.

Russ, Thomas H. 2000. *Brownfield Redevelopment.* New York: McGraw-Hill.

Sustainable Chattanooga. 1995. *Cleaning Up Chattanooga Creek.* Chattanooga, Tenn.: President's Council on Sustainable Development Briefing Book. Found at: <http://www.chattanooga.net/sustain/pcsd_briefing_book/resources_chat_creek.html>.

USEPA. 1998. Office of Solid Waste and Emergency Response. *Characteristics of Sustainable Brownfields Projects.* EPA500-R-98-001. Washington, D.C.: USEPA.

USEPA. 1999. Office of Solid Waste and Emergency Response. *A Sustainable Brownfields Model Framework.* EPA500-R-99-001. Washington, D.C.: USEPA.

USEPA. 2002. *Brownfields: A Definition.* Washington, D.C.: U. S. Environmental Protection Agency, Office of Solid Waste and Emergency Response. Found at: <http://www.epa.gov/swerosps/bf/glossary.htm#brow>.

Zachary, Jill. 1995. *Sustainable Community Indicators: Guideposts for Local Planning.* Santa Barbara, Calif.: Community Environmental Council, Inc.

Chapter 4

About Smart Growth. 2001. Washington, D.C.: Smart Growth Network, Sustainable Communities Network. Found at: <http://www.smartgrowth.org/about/default.asp>.

Anderson, Geoff. 1998. *Why Smart Growth: A Primer.* Washington, D.C.: International City/County Management Association.

Beatley, Timothy, and Kristie Manning. 1997. *The Ecology of Place.* Washington, D.C.: Island Press.

Breslaw, Jon A. 1990. Density and Urban Sprawl: Comment. *Land Economics* 66, no. 43: 464–468.

Brueckner, Jan K. 1995. *Infrastructure Financing and Urban Development: The Economics of Impact Fees.* Urbana-Champaign: Institute of Government and Public Affairs, University of Illinois at Urbana-Champaign, October.

Charles, John A. 1998, The Dark Side of Growth Controls: Some Lessons from Oregon. Pittsburgh, Pa.: The Allegheny Institute. Found at: <http://www.alleghenyinstitute.org/reports.php>.

Chattanooga Chamber of Commerce. 2001. CRGI: A Progress Report of the Chattanooga Regional Growth Initiative. Chattanooga Area Chamber of Commerce. Found at: <http://www.chattanooga-chamber.com/porter/art/CRGI_report.pdf>.

Chattanooga Chamber of Commerce. 2000. Former Army Land Transferred to City and County Ownership. Chattanooga, Tenn.: Chattanooga Chamber of Commerce. Found at: <http://chattanooga-chamber.com/newsandinfo/headline.php3?id=36>.

City of Chattanooga. 1995. *President's Council on Sustainable Development Briefing Book.* Found at: <http://www.chattanooga.net/sustain/pcsd_briefing_book/industry_eco_industry.html>.

City of Olympia. 2001. *The Sustainable Community Roundtable.* Olympia, Wash.: The Sustainability Roundtable, Inc. Found at: <http://www.olywa.net/roundtable>.

City of Portland. 2000. *Chapter 33.430: Environmental Zones.* Portland, Ore.: Bureau of Planning. Found at: <http://www.planning.ci.portland.or.us/zoning/ZCtest/400/430_Envir.pdf>.

City of Santa Monica. 2002. *Santa Monica Green Building Design and Construction Guidelines: Introduction.* Santa Monica, Calif.: Green Building Program. Found at: <http://greenbuildings.santa-monica.org/>.

City of Scottsdale. 2001. *Scottsdale's Green Building Program: Sustainable Living in the Desert.* Scottsdale, Ariz.: Green Building Program. Found at: <http://www.ci.scottsdale.az.us/greenbuilding/>.

City of Seattle. 2002. *Seattle Comprehensive Plan, Land Use Element.* Seattle, Wash.: Strategic Planning Office, Department of Comprehensive and Neighborhood Planning. Found at: <http://www.cityofseattle.net/planning/comprehensive/HomeCP.htm>.

Cornell University. 2001. *Eco-Industrial Development.* Ithaca, N.Y.: Center for the Environment, Work and Environment Initiative. Found at: <http://www.cfe.cornell.edu/wei/EID.html>.

Downs, Anthony. 2000. *Dealing Effectively with Fast Growth.* Washington, D.C.: Brookings Institution, Policy Brief No. 67.

English, Mary P., Jean H. Peretz, and Melissa Mandershield. 1999. *Smart Growth for Tennessee Towns & Cities: A Process Guide.* Knoxville Tenn.: Waste Management Research and Education Institute. Found at: <http://eerc.ra.utk.edu/smart/title.htm>.

Fischel, W. A. 1990. *Do Growth Controls Matter: A Review of Empirical Evidence on the Effectiveness and Efficiency of Local Government Land Use Regulation.* Cambridge, Mass.: Lincoln Institute of Land Policy.

Franciosi, Robert. 1998. A Tale of Two Cities: Phoenix, Portland, Growth and Growth Control. Arizona Issue Analysis 152, Phoenix, Ariz.: The Goldwater Institute, October. Comparative analysis of growth controls in Portland, Oregon, and Phoenix, Arizona. Available at: <http://goldwaterinstitute.org/>.

Frieden, Bernard J. 1979. *The Environmental Protection Hustle*. Cambridge, Mass.: MIT Press.

Goode, Ann Eberhart, Elizabeth Collaton, and Charles Bartsch. 2000. *Smart Growth*. Washington, D.C.: Northeast-Midwest Institution, April. Found at: <http://www.nemw.org/ERsmartgrowth.htm>.

Gordon, Peter, and Harry W. Richardson. 1998. Prove It: The Costs and Benefits of Urban Sprawl. *The Brookings Review* 16, no. 4: 23–25. Found at: <http://www.brook.edu/press/review/fa98/gordon2.pdf>.

Hogen-Esch, Thomas. 2001. Washington's Growth Management Act and the Neighborhood Movement in West Seattle. Paper delivered at the Meetings of the American Political Science Association, San Francisco, Calif., August 30–September 2, 2001.

Iglitzin, Lynne B. 1995. The Seattle Commons: A Case Study of the Politics and Planning of an Urban Village. *Policy Studies Journal* 23: 620–633.

Judd, Dennis R., and Todd Swanstrom. 1998. *City Politics: Private Power and Public Policy*. New York: Longman.

Kline, Elizabeth. 1995. *Sustainable Community Indicators*. Medford, Mass.: Consortium for Regional Sustainability, Tufts University.

Leo, Christopher. 1998. Regional Growth Management Regime: The Case of Portland, Oregon. *Journal of Urban Affairs* 20: 363–394.

Molotch, Harvey. 1976. The City as Growth Machine: Toward a Political Economy of Place. *American Journal of Sociology* 82, no. 2: 309–332.

Myers, Phyllis. 1999. Livability at the Ballot Box: State and Local Referenda on Parks, Conservation, and Smarter Growth, Election Day 1998. Paper prepared for the Brookings Institution Center on Urban and Metropolitan Policy. Washington, D.C.: Brookings Institution.

Myers, Phyllis, and Puentes, Robert. 2001. Growth at the Ballot Box: Electing the Shape of Communities in November 2000. Paper prepared for the Brookings Institution Center on Urban and Metropolitan Policy. Washington, D.C.: Brookings Institution.

Nelson, Arthur C., and Terry Moore. 1996. Assessing Growth Management Policy Implementation: Case Study of the United States' Leading Growth Management State. *Land Use Policy* 13, no. 4: 241–259.

PACE. 2000. *Partners for a Clean Environment Program Overview*. Boulder, Colo.: City of Boulder. Found at: <http://www.ci.boulder.co.us/environmentalaffairs/PACE/business_menu.html>.

Peiser, Richard B. 1989. Density and Urban Sprawl. *Land Economics* 65, no. 3: 194–204.

Smart Growth Network. 2000a. *Eco-Industrial Case Studies.* Washington, D.C.: International City/County Management Association. Found at: <http://www.smartgrowth.org/library/eco_ind_case_intro.html> and <http://www.smartgrowth.org/casestudies/casestudy_index.html>.

Smart Growth Network. 2000b. *Case Studies: The Volunteer Site, Chattanooga, TN.* Washington, D.C.: International City/County Management Association. Found at: <http://www.smartgrowth.org/casestudies/ecoin_chattanooga.html>.

Smarth Growth Network. 2001. *Smart Growth State by State.* Washington, D.C.: International City/County Management Association. Found at: <http://www.smartgrowth.org/news>.

Spohn, Suzanne G. 1997. Eco-Industrial Parks Offer Sustainable Base Redevelopment. *International City/County Management Association Base Reuse Consortium Bulletin*, May, no pages given. Found at: <http://www.smartgrowth.org/library/spohn_icma.html>.

Stone, Clarence. 1993. Urban Regimes and the Capacity to Govern. *Journal of Urban Affairs* 15, no. 1, 1–28.

Washington State. 1990. Washington State Growth Strategies Commission. *A Growth Strategy for Washington State—Final Report.* Seattle, Wash.: State Department of Community Development, September.

Young, Dwight. 1995. *Alternatives to Sprawl.* Cambridge, Mass.: Lincoln Institute of Land Policy.

Chapter 5

Agyeman, Julian, and Bob Evans. 1995. Sustainability and Democracy: Community Participation in Local Agenda 21. *Local Government Policy Making* 22, no. 2: 35–40.

AtKisson, Alan. 1999. Developing Indicators of Sustainable Community: Lessons from Sustainable Seattle. In *The Earthscan Reader in Sustainable Cities*, David Satterthwaite (ed.), pp. 352–363. London: Earthscan.

Barber, Benjamin R. 1984. *Strong Democracy: Participatory Politics for a New Age.* Berkeley: University of California Press.

Berry, Jeffrey, Kent E. Portney, and Ken Thomson. 1993. *The Rebirth of Urban Democracy.* Washington, D.C.: Brookings Institution Press.

Boston. 2000. Interview with Charlotte Kahn, The Boston Foundation, April 28.

Brugmann, Jeb. 1997. Sustainability Indicators Revisited: Getting from Political Objectives to Performance Outcomes—A Response to Graham Pinfield. *Local Environment* 2, no. 3: 299–302.

Buell, John, and Tom DeLuca. 1996. *Sustainable Democracy: Individuality and the Politics of the Environment.* Thousand Oaks, Calif.: Sage.

City of Seattle. 2002. *Seattle's Comprehensive Plan.* Seattle, Wash.: Strategic Planning Office, Department of Comprehensive and Neighborhood Planning. Found at: <http://www.cityofseattle.net/planning/comprehensive/HomeCP.htm>.

Cook, Fay Lomax, and Lawrence R. Jacobs. n.d. *Deliberative Democracy in Action: Evaluation of Americans Discuss Social Security*. Report to the Pew Charitable Trusts. Evanston, Ill.: Institute for Policy Research, Northwestern University.

Cook, Fay Lomax, Jason Barabas, and Lawrence R. Jacobs. 1999. Deliberative Democracy in Action: An Analysis of the Effects of Public Deliberation. Working Paper. Institute for Policy Research, Northwestern University, Evanston, Ill.

Daly, Herman, and John Cobb. 1994. *For the Common Good: Redirecting the Economy toward Community, the Environment, and a Sustainable Future*. New York: Beacon Press.

Dryzek, John S. 1987. *Rational Ecology: Environment and Political Economy*. New York: Basil Blackwell.

Etzioni, Amatai. 1996. *The New Golden Rule: Community and Morality in a Democratic Society*. New York: Basic Books.

Hardin, Garrett. 1968. The Tragedy of the Commons. *Science* 162: 1243–1248.

Ingram, Helen, Lenard Milich, and Robert G. Varady. 1994. Managing Transboundary Resources: Lessons from Ambos Nogales. *Environment* 36, no. 4: 6–9, 28–38.

Jacksonville Community Council, Inc. 2000. *Quality of Life Indicators in Jacksonville: Indicators for Progress*. Jacksonville, Fla.: Jacksonville Community Council, Inc.

John, DeWitt. 1994. *Civic Environmentalism: Alternatives to Regulation in States and Communities*. Washington, D.C.: Congressional Quarterly Press.

John, DeWitt, and M. Mlay. 1997. Community-Based Environmental Protection: Encouraging Civic Environmentalism. In *Better Environmental Decisions: Strategies for Governments, Businesses, and Communities*, in K. Sexton, A. Marucs, K. Easter, and T. Burkhardt (eds.), pp. 353–376. Washington, D.C.: Island Press.

Mazmanian, Daniel, Michael Stanley-Jones, and Miriam Green. 1988. *Breaking Political Gridlock: California's Experience in Public-Private Cooperation for Hazardous Waste Policy*. Claremont, Calif.: California Institute of Public Affairs.

National Commission on Civic Renewal. 1998. *A Nation of Spectators: How Civic Disengagement Weakens America and What We Can Do about It*. College Park, Md.: National Commission on Civic Renewal.

Olympia. 2000. Interview with Dorothy Craig, co-founder of Olympia's Sustainability Roundtable, April 17.

Ophuls, William, and A. Stephen Boyan. 1992. *Ecology and the Politics of Scarcity Revisited: The Unraveling of the American Dream*. New York: W. H. Freeman.

Ostrom, Elinor. 1990. *Governing the Commons: The Evolution of Institutions for Collective Action*. New York: Cambridge University Press.

Ostrom, Elinor, Roy Gardner, and James Walker. 1994. *Rules, Games, and Common-Pool Resources*. Ann Arbor: University of Michigan Press.

Pallemaerts, Marc. 1988. The Politics of Acid Rain Control in Europe. *Environment* 31, no. 2: 42–44.

Portney, Kent E. 1992. *Siting Hazardous Waste Treatment Facilities: The NIMBY Syndrome.* Westport, Conn.: Auburn House, Greenwood Publishing Group.

Potapchuk, William. 1996. Building Sustainable Community Politics: Synergizing Participatory, Institutional, and Representative Democracy. *National Civic Review* 85, no. 3: 54–59.

Press, Daniel. 1994. *Democratic Dilemmas in the Age of Ecology: Trees and Toxics in the American West.* Durham, N.C.: Duke University Press.

Putnam, Robert. 1995. Bowling Alone: America's Declining Social Capital. *Journal of Democracy* 6: 65–78.

Putnam, Robert. 2000. *Bowling Alone: The Collapse and Revival of American Community.* New York: Simon and Schuster.

Rabe, Barry G. 1992. When Siting Works, Canada Style. *Journal of Health Politics, Policy, and Law* 17: 119–142.

Sandel, Michael J. 1984. The Procedural Republic and the Unencumbered Self. *Political Theory* 12, no. 1: 81–96.

Santa Barbara. 1998. *Santa Barbara South Coast Community Indicators.* Santa Barbara, Calif.: Community Environmental Council. Found at: <http://www.library.ucsb.edu/docs/indicators98.pdf>.

Selman, Paul, and Jane Parker. 1997. Citizenship, Civicness, and Social Capital in Local Agenda 21. *Local Environment* 2, no. 2: 171–187.

Stone, Clarence N. 1993. Urban Regimes and the Capacity to Govern. *Journal of Urban Affairs* 15, no. 1: 1–28.

Switzer, Jacqueline Vaughn. 1997. *Green Backlash: The History and Politics of Environmental Opposition in the U.S.* Boulder, Colo.: Lynne Reiner.

Urahn, Susan K., and Marshall A. Ledger. 2000. Where Engagement Starts: The Americans Discuss Social Security Project. *Trust* (April). Found at: <http://www.pewcharitabletrusts.org/ideas/ideas_item.cfm?content_item_id=345&content_type_id=11&issue_name=&issue=6&page=11&name=Lessons%20Learned%20%28%3Ci%3ETrust%3C%2Fi%3E%29>.

Walsh, Edward J., Rex Warland, and D. Clayton Smith. 1997. *Don't Burn It Here: Grassroots Challenges to Trash Incinerators.* University Park, Pa.: Pennsylvania State University Press.

Williams, Bruce A., and Albert R. Matheny. 1995. *Democracy, Dialogue, and Environmental Disputes: The Contested Languages of Social Regulation.* New Haven, Conn.: Yale University Press.

Young, Oran. 1994. *International Governance: Protecting the Environment in a Stateless Society.* Ithaca, N.Y.: Cornell University Press.

Zachary, Jill. 1995. *Sustainable Community Indicators: Guideposts for Local Planning.* Santa Barbara, Calif.: Community Environmental Council.

Chapter 6

Anderton, Douglas L., Andy B. Anderson, Peter H. Rossi, John Michael Oakes, Michael R. Fraser, Eleanor W. Weber, and Edward J. Calabrese. 1994. Hazardous Waste Facilities: "Environmental Equity" Issues in Metropolitan Areas. *Evaluation Review* 18, no. 2: 123–140.

AtKisson, Alan. 1999. Developing Indicators of Sustainable Community: Lessons from Sustainable Seattle. In *The Earthscan Reader in Sustainable Cities*. David Satterthwaite (ed.), pp. 352–363. London: Earthscan.

Bryant, Bunyan, and Paul Mohai. 1990. *The Proceedings of the Michigan Conference on Race and the Incidence of Environmental Hazards*. Ann Arbor.: University of Michigan School of Natural Resources and Environment.

Bryant, Bunyan, and Paul Mohai, eds. 1992. *Race and the Incidence of Environmental Hazards: A Time for Discourse*. Boulder, Colo.: Westview Press.

Bullard, Robert. 1990. *Dumping in Dixie: Race, Class, and Environmental Quality*, Boulder, Colo.: Westview Press.

Bullard, Robert, ed. 1994. *Unequal Protection: Environmental Justice and Communities of Color*. San Francisco, Calif.: Sierra Club.

Camacho, David E., ed. 1998. *Environmental Injustices, Political Struggles: Race, Class, and the Environment*. Durham, N.C.: Duke University Press.

City of Austin. 2000. *CIP Sustainability Matrix*. Austin, Texas: Department of Planning, Environmental, and Conservation Services. Found at: <http://www.ci.austin.tx.us/sustainable/matrix.htm> and <http://www.ci.austin.tx.us/sustainable/matrixintro.htm>.

City of San Francisco. 1996. *Sustainability Plan for the City of San Francisco, Indicators and Strategy*. San Francisco, Calif.: SFEnvironment. Found at: <http://www.sfgov.org/sfenvironment/pages/sup_pages/homepage_items/justice/main.htm> and <http://www.sustainable-city.org/Plan/Justice/strategy.htm>.

City of Seattle. 2002. *Seattle's Comprehensive Plan*. Seattle, Wash.: Strategic Planning Office, Department of Comprehensive and Neighborhood Planning. Found at: <http://www.cityofseattle.net/planning/comprehensive/HomeCP.htm>.

Dobson, Andrew. 1998. *Justice and the Environment: Conceptions of Environmental Sustainability and Theories of Distributive Justice*. New York: Oxford University Press.

Dore, Mohammad, and Timothy Mount, eds. 1999. *Global Environmental Economics: Equity and the Limits of Markets*. Malden, Mass.: Blackwell.

Foreman, Christopher H. 1998. *The Promise and Peril of Environmental Justice*. Washington, D.C.: Brookings Institution Press.

Frieden, Bernard J. 1979. *The Environmental Protection Hustle*. Cambridge, Mass.: MIT Press.

Geiser, K., and G. Waneck. 1994. PCBs and Warren County. In *Unequal Protection: Environmental Justice and Communities of Color*, Robert Bullard (ed.), pp. 43–52. San Francisco, Calif.: Sierra Club.

Global Cities Online. 2000. *Indicators of Sustainability.* Costs and Benefits, Social Equity, and Poverty. Found at: <http://www.globalcities.org/intropg/demo/PGI/indicat.html#Costs and Benefits>.

Goldman, Benjamin A. 1993. *Not Just Prosperity: Achieving Sustainability with Environmental Justice.* Vienna, Va.: National Wildlife Federation.

Haughton, G. 1999. Environmental Justice and the Sustainable City. *Journal of Planning Education and Research* 18, no. 3: 233–243.

Jacksonville Community Council, Inc. 1999. *Quality of Life in Jacksonville: Indicators for Progress.* Jacksonville, Fla.: Jacksonville Community Council, Inc. Found at: <http://www.jcci.org/qol/qol.htm>.

Kee, Warner, and Harvey Molotch. 2000. *Building Rules: How Local Controls Shape Community Environments and Economics.* Boulder, Colo.: Westview Press.

Kline, Elizabeth. 1995a. *Sustainable Community Indicators.* Medford, Mass.: Consortium for Regional Sustainability, Tufts University.

Kline, Elizabeth. 1995b. *Sustainable Community Indicators: Examples from Cambridge, MA.* Medford, Mass.: Consortium for Regional Sustainability, Tufts University.

Lewis, Paul, and Max Neiman. 2001. Local Governments, Housing, and Growth Management: Balancing Act or "Hustle"? Paper delivered at the 2001 Meetings of the American Political Science Association, San Francisco, August 30–September 2, 2001.

McGranahan, Gordon, Jacob Singsore, and Marianne Kjellen. 1996. Sustainability, Poverty and Urban Environmental Transitions. In *Sustainability, the Environment, and Urbanization*, Gordon McGranahan, Jacob Singsore, and Marianne Kjellen (eds.), pp. 103–134. Washington, D.C.: Earthscan.

Mohai, Paul, and Bunyan Bryant. 1992a. Environmental Injustice: Weighing Race and Class as Factors in the Distribution of Environmental Hazards. *University of Colorado Law Review* 63, no. 4: 921–932.

Mohai, Paul, and Bunyan Bryant. 1992b. Race, Poverty, and the Environment. *EPA Journal* 18, no. 1: 6–8.

Neiman, Max, and Ronald O. Loveridge. 1981. Environmentalism and Local Growth Control: A Probe into the Class Bias Thesis. *Environment and Behavior* 13: 759–772.

Perkins, Jane. 1992. Recognizing and Attacking Environmental Racism. *Clearinghouse Review* 26: 389.

Portney, Kent E. 1991a. *Siting Hazardous Waste Treatment Facilities: The NIMBY Syndrome.* Westport, Conn.: Auburn House, Greenwood Publishing Group.

Portney, Kent E. 1991b. Public Environmental Policy Decision Making: Citizen Roles. In *Environmental Decision Making: A Multidisciplinary Perspective*, Richard A. Chechile and Susan Carlisle (eds.), pp. 195–215. New York: Van Nostrand.

Portney, Kent E. 1994. Environmental Justice and Sustainability: Is There a Critical Nexus in the Case of Hazardous Waste Disposal or Treatment Facility Siting? *Fordham Urban Law Journal* 21, no. 3: 827–839.

Rees, Matthew. 1992. Black and Green: The Birth of "Eco-Racism." *The New Republic* 206, no. 9: 15–16.

Robinson, John, George Francis, Russel Legge, and Sally Lerner. 1990. Defining a Sustainable Society: Values, Principles, and Definitions. *Alternatives* 17, no. 2: 36–46.

Sustainability Indicators Project of Hays, Travis, and Williamson Counties. 2000. *Central Texas Indicators 2000: A Report on the Economic, Environmental, and Social Health of the Central Texas Region.* Found at: <http://www.centex-indicators.org/pubs/centex-indicators2000.pdf>.

UCC. 1987. United Church of Christ, Commission for Racial Justice. *Toxic Wastes and Race in the United States.* New York: United Church of Christ.

USEPA. 1992. *Environmental Equity: Reducing the Risks.* Washington, D.C.: National Service Center for Environmental Publications, February.

USGAO. 1983. U.S. Government Accounting Office. *Siting of Hazardous Waste Landfills and their Correlation with Racial and Economic Status of Surrounding Communities.* Washington, D.C.: U.S. Government Printing Office.

Walker, Nathalie and Michael Traynor. 1992. The Environmental Justice Movement: Two Cases in Point. *Environmental Law* 12: 3–4.

Walsh, Edward J., Rex Warland, and D. Clayton Smith. 1997. *Don't Burn It Here: Grassroots Challenges to Trash Incinerators.* University Park, Pa.: Pennsylvania State University Press.

WCED. (World Commission on Environment and Development—The Brundtland Commission). 1987. *Our Common Future.* New York: Oxford University Press.

Chapter 7

AtKisson, Alan. 1996. Developing Indicators of Sustainable Community: Lessons from Sustainable Seattle. *Environmental Impact Assessment Review* 16: 337–350.

Austin Energy. 2001. *GreenChoice: The Power to Choose.* Austin, Texas: Austin Energy. Found at: <http://www.AustinEnergy.com/greenchoice/>.

Besleme, Kate, Elisa Maser, and Judith Silverstein. 1999. *A Community Indicators Case Study: Addressing the Quality of Life in Two Communities.* San Francisco, Calif.: Redefining Progress.

Boulder City Council. 2001. City of Boulder 2000–2001 City Council Goals: Environmental Sustainability Update, One Year Status Report, January 20. Boulder, Colo.: City of Boulder. Found at: <http://www.ci.boulder.co.us/cmo/citycouncil/goals1-01/environ4.html>.

Brugmann, Jeb. 1997. Sustainability Indicators Revisited: Getting from Political Objectives to Performance Outcomes. *Local Environment* 2, no. 3: 299–302.

Central Texas Indicators 2001. 2001. *Sustainability Indicators Project of Bastrop, Caldwell, Hays, Travis, and Williamson Counties: Annual Report on the Economic, Environmental, and Social Health of the Central Texas Region.* Austin, Tex. Found at: <http://www.centex-indicators.org/ar2001.html>.

Charles, John A. 1998. *The Dark Side of Growth Controls: Some Lessons from Oregon.* Pittsburgh, Pa.: The Allegheny Institute. Found at: <http://www.alleghenyinstitute.org/reports.php>.

Chattanooga Area Chamber of Commerce. 2000. Interview with Rich Bailey, Media Relations Specialist, Chattanooga Chamber of Commerce, May 1, 2000.

CHCRPA. 1997. Chattanooga Hamilton County Regional Planning Agency. *Futurescape Survey Results.* Chattanooga, Tenn.: CHCRPA.

City of Austin. 1999. *The CIP Sustainablility Matrix.* Austin, Texas: City of Austin. Found at: <http://www.ci.austin.tx.us/sustainable/matrix.htm>.

City of Austin. 2000. *Smart Growth Initiative: Traditional Neighborhood Development (TND) and Transit-Oriented Development (TOD).* Austin, Texas: City of Austin. Found at: <http://www.ci.austin.tx.us/smartgrowth/tnd.htm>.

City of Austin. 2001a. *Green Building Program: Fostering Green Building through Local Government Initiatives.* Austin, Texas: City of Austin. Found at: <http://www.ci.austin.tx.us/greenbuilder/suepaper.htm>.

City of Austin. 2001b. *Resources for Choosing Products and Services That Meet Sustainability Criteria.* Austin, Texas: City of Austin. Found at: <http://www.ci.austin.tx.us/sustainable/purchasing.htm>.

City of Boulder. 2002. PACE Program Overview. Boulder, Colo.: City of Boulder, Department of Environmental Affairs. Found at: <http://www.ci.boulder.co.us/environmentalaffairs/PACE/index.htm>.

City of Jacksonville. 2000. *Quality of Life in Jacksonville: Indicators of Progress.* Jacksonville, Fla: Jacksonville Community Council, Inc. Found at: <http://www.jcci.org/qol/qol.htm>.

City of Portland. 1999. *Comprehensive Plan Goals and Policies,* revised. Portland, Ore.: City of Portland. Found at: <http://www.planning.ci.portland.or.us/pdf/ComprehensivePlan.pdf>.

City of San Francisco. n.d. *The Sustainability Plan for the City and County of San Francisco: A Brief Overview.* Found at: <http://www.sfgov.org/sfenvironment/environment/sustain/excerpt.htm>.

City of San Francisco. 1996. *The Sustainability Plan for the City and County of San Francisco.* October. San Francisco, Calif.: City of San Francisco. Found at: <http://www.sfgov.org/sfenvironment/environment/sustain/index.htm>.

City of Santa Monica. 1996. *Sustainable City Progress Report.* Santa Monica, Calif.: Task Force on the Environment. December. Found at: <http://pen.ci.santa-monica.ca.us/environment/policy/progress4.pdf>.

City of Santa Monica. 1999. *Sustainable City Progress Report*. Santa Monica, Calif.: Task Force on the Environment. Found at: <http://pen.ci.santamonica.ca.us/environment/policy/SCPRU99Full.PDF>.

City of Seattle. 2000a. *Seattle's Comprehensive Plan*. Seattle, Wash.: Strategic Planning Office: Found at: <http://www.ci.seattle.wa.us/planning/CompPlan/CP1-LandUse.htm>.

City of Seattle. 2000b. *Environmental Management Program*. Seattle, Wash.: Office of Sustainability and Environment. Found at: <http://www.ci.seattle.wa.us/oem/EMPIndex.htm>.

Graham, Lamar. 1999. The Reborn American City: A Place Where You Might Want to Live. *Parade* (April 25): 4–6.

Magilavy, Beryl. 1998. Indicators Applications: Moving Indicators into Action, San Francisco's Experience 1988–1998. Paper presented at the 1998 Redefining Progress California Community Indicators Conference, San Francisco, December 4 Found at: <http://www.sustainable-city.org/document/assess.htm>.

Metro Council. 1998. *Life in Hamilton County, 1998: Indicators of Community Well-Being*. Chattanooga, Tenn.: Metropolitan Council for Community Services, Inc., November 10. Found at: <http://www.chattanooga.net/metrocouncil/indicators%20pages/tableofcontents.html>.

Parr, John. 1998. Chattanooga: The Sustainable City. In *Boundary Crossers: Case Studies of How Ten of America's Metropolitan Regions Work*, Bruce Adams and John Parr (eds.), pp. 68–80. College Park, Md.: Academy of Leadership.

Pierce, Neal, and Curtis Johnson. 1998. A Civic Vignette: The Chattanooga Story—From Troubled Raw River Town to Global Model. In *Boundary Crossers: Community Leadership for a Global Age*, Neal Pierce and Curtis Johnson (eds.). College Park, Md.: Academy of Leadership.

Portland Office of Sustainable Development. 2002. *City Scorecard*. Portland, Ore.: Office of Sustainable Development. Found at: <http://www.sustainableportland.org/Scorecard.pdf>.

Redefining Progress. 1999. Proceedings of the California Community Indicators Conference. San Francisco, Calif.: Redefining Progress. Found at: <http://www.rprogress.org/pubs/pdf/cip_dec98_proc.pdf>.

San Francisco Commission on the Environment. 2002. *Reports and Resolutions*. San Francisco, Calif.: Commission on the Environment. Found at: <http://www.sfgov.org/sfenvironment/pages/sup_pages/envir_legislation/resolution/envir_resolutions.htm>.

Skinnarland, Kirvil. 1999. Thinking Globally, Acting Locally: The Role of Cities in Sustainable Development—A Case Study of the City of Seattle. Paper presented at the 1999 meetings of the Greening of Industry Network, Chapel Hill, N.C.

Sustainable Seattle. 2000a. *Sustainable Seattle: About Us*. Seattle. Wash.: Sustainable Seattle Inc. Found at: <http://www.scn.org/sustainable/about.htm>.

Sustainable Seattle. 2000b. Sustainable Seattle and the Challenge of Sustainability. Seattle, Wash.: Sustainable Seattle, Inc. Found at: <http://www.scn.org/sustainable/sust-challenge.htm>.

Chapter 8

Beach, David. 2000. Progress in the Cleveland EcoVillage. *Ecocity Cleveland Journal* 7, nos. 10–11. Found at: <http://www.ecocleveland.org/archive/html/nov2000/progress.htm>.

Boston Foundation. 2000. *The Wisdom of our Choices: Boston's Indicators of Progress, Change, and Sustainability*. Boston, Mass.: The Boston Foundation. Summary found at: <http://www.tbf.org/boston/index.html>.

Boston Redevelopment Authority (BRA). 2000. Interview with Robert Consalvo, Director of Research, Boston Redevelopment Authority, April 28, 2000.

City of Boston. 2000. *Sustainable Boston Initiative*. Boston, Mass.: City of Boston. Found at: <http://www.ci.boston.ma.us/environment/sustain.asp>.

City of Cambridge. 1993. *Toward a Sustainable Future*. Cambridge, Mass.: Cambridge Planning Board and Community Development Department.

City of Cambridge. 2001a. *Citywide Growth Management Advisory Committee*. Cambridge, Mass.: City of Cambridge. Found at: <http://www.ci.cambridge.ma.us/~CDD/commplan/zoning/growthcomm/>.

City of Cambridge. 2001b. *Citywide Rezoning Petition as Passed February 2001*. Cambridge, Mass.: City of Cambridge. Found at: <http://www.ci.cambridge.ma.us/~CDD/commplan/zoning/cityrezoneprop/index.html>.

City of San Jose. 1998a. *San Jose Sustainable City Status Report, Executive Summary*. San Jose, Calif.: Environmental Services Department. Found at: <http://www.ci.san-jose.ca.us/esd/suscityexecsummary.htm>.

City of San Jose. 1998b. *San Jose's Sustainable Cities Programs, Draft Status Report, June 1998*. San Jose, Calif.: Environmental Services Department. Found at: <http://www.ci.san-jose.ca.us/esd/Word-Excel/SCRPTFT.DOC>.

City of Tampa. 1999. *Sustainable Communities Second Annual Report*. July 1999. Tampa, Fla.: City of Tampa, Department of Planning. Found at: <http://www.tampagov.net/dept_Planning/planning_section/Sustainable/index.asp>.

City of Tampa. 2001. *Sustainable Communities Designation Agreement*. Tampa, Fla.: City of Tampa, Department of Planning. Found at: <http://www.tampagov.net/dept_Planning/planning_section/Sustainable/agreement.htm>.

Cleveland EcoVillage. 2001. *Cleveland EcoVillage: Introduction*. Cleveland, Ohio: City of Cleveland. Found at: <http://www.ecocleveland.org/ecovillage/index.html>.

Craig, Dorothy. 2000. Telephone interview conducted with Dorothy Craig, co-founder of *The Sustainability Roundtable*, April 17, 2000.

DeLeon, Richard E. 1992. *Left Coast City: Progressive Politics in San Francisco, 1975–1991*. Lawrence, Kans.: University of Kansas Press.

Downs, Anthony. 2000. *Dealing Effectively with Fast Growth*. Washington, D.C.: Brookings Institution Center on Urban and Metropolitan Policy, Policy Brief No. 67, November. Available at: <http://www.brook.edu/dybdocroot/comm/policybriefs/pb067/pb67.htm>.

EcoCity Cleveland. 2001. *EcoCity Cleveland: Ecological Thinking and Environmental Planning for Northeast Ohio—The Cuyahoga Bioregion*. Cleveland, Ohio: EcoCity Cleveland. Found at: <http://www.ecocleveland.org/>.

Esty, Daniel C. 2001. *2001 Environmental Sustainability Index: An Initiative of the Global Leaders of Tomorrow Environmental Task Force*. New York: World Economic Forum. Available at: <http://www.ciesin.columbia.edu/indicators/ESI>.

NEO-EMPACT. 2001. Northeast Ohio EMPACT. Cleveland, Ohio: NEO-EMPACT. Found at: <http://empact.nhlink.net/docs/neoframe.html>.

San Jose Planning Department. 1999. Interview with Laurel Prevetti, Principal Planner, San Jose Planning Department, April 28. 1999.

State of Florida. 2001. *Florida Sustainable Communities Demonstration Project*. Tampa, Fla.: City of Tampa, Department of Planning. Found at: <http://www.tampa.gov.net/dept_Planning/planning_section/Sustainable/contract.htm>.

Stone, Clarence N. 1993. Urban Regimes and the Capacity to Govern. *Journal of Urban Affairs* 15, no. 1: 1–28.

Town of Brookline. 2001a. *Town of Brookline Comprehensive Plan Pages*. Brookline, Mass.: Town of Brookline, Department of Planning. Found at: <http://www.townofbrooklinemass.com/Planning/ComprehensivePlan/techreports.htm>.

Town of Brookline. 2001b. *Sustainability: Issues and Opportunities*. Brookline, Mass.: Town of Brookline, Department of Planning. Found at: <http://www.townofbrooklinemass.com/Planning/ComprehensivePlan/SustainabilityWEB.PDF>.

World Resources Institute. 1993. *The 1993 Information Please Environmental Almanac*. Boston, Mass.: Houghton Mifflin.

Index

Accra, Ghana, 160
Action plans, in Boulder, 206
Advanced Vehicle Systems, Inc. (AVS), 82, 99, 110, 186
Affirmative action, 27
Affordable housing, 12, 27
Agency for Toxic Substances and Disease Registry, 84
Agenda 21, 10, 14
Agenda-setting processes, 32
Agyeman, Julian, 134
Air pollution abatement, in Chattanooga, 32
Air pollution reduction, 67
Air quality, 78, 79–82
Alberta, Canada, 148
Alliance for Sustainable Community (Annapolis), 23
Alton Park (Chattanooga), 93
Americans Discuss Social Security, 147
Anderson, Geoff, 105
Anderton, Douglas L., 164
Annapolis, Md., 64
Annapurna Conservation Cooperative Project (Nepal), 12
Antigrowth, relation to smart growth, 9
Aquifer protection, 79
Arizona, State of, 249
Arizona, comprehensive planning requirements, 106
Asbestos abatement, 67

Asbestos, disparities in exposure to, 162
AtKisson, Alan, 66, 152, 170, 194, 213
Atlanta, Georgia, 26, 115
Austin Energy, Inc., 97, 184
Austin, Tex., 22, 25, 29, 40, 56, 68, 69, 86, 87, 96, 97, 110, 147, 169, 172, 173, 174, 175, 178, 179, 181–185, 208, 217, 219, 243, 244, 247
  Department of Planning, Environmental, and Conservation Services, 63
  light rail ballot question in, 107
Auto body shops, in Boulder, 207

Bald, Andre, 19
Ball Aerospace and Technologies, 203
Baltimore, Md., 91, 92, 115
Baltimore Brownfields Project, 91
Baltimore Development Corporation, 91
Bank of America, 19
Barabas, Jason, 147
Barber, Benjamin R., 132
Barton, Hugh, 12
Bartsch, Charles, 105
Battery-powered electric vehicles, 110
Beatley, Timothy, 1, 12, 18, 21, 37, 118
Bell, Colin, 11
Benchmarks, 40, 66

Berry, Jeffrey M., 15, 147
Besleme, Kate, 214
Best practices in cities, 177
Bicycle ridership program, 67
Biodiversity, 94
Biophysical environment and ecology, 28
Biosphere, elements of sustainability, 4
Blue Blazes Historic Trail (Tenn.), 188
Boston, Mass., 22, 24, 40, 68, 69, 99, 154, 222, 223, 224
Boston Foundation, 224
Boston Redevelopment Authority (BRA), 225
Boulder, Colo., 22, 29, 67, 69, 79, 81, 82, 88, 97, 99, 117, 120, 178, 179, 190, 202–207, 217, 219, 228, 244, 247
Boulder Chamber of Commerce, 81, 117, 206
Boulder County, 205
Boulder County Clean Air Consortium, 81, 203
Boulder County Health Department, 81, 117, 206
Boulder Creek, 205
Boulder Creek Corridor Project, 205
Boulder, Department of Environmental Affairs, 63
Boulder Energy Conservation Center, 81, 117, 206
Boulder Greenways program, 205
Boulder Office of Environmental Affairs, 81, 117
Boulder Valley Comprehensive Plan, 203
Boulder Water Quality and Environmental Services, 117
Boyan, A. Stephen, 94, 128, 135, 136, 144
Brainerd Levee, 188
Breslaw, Jon A., 106
Brett, Deborah, 19
Bridge to a Sustainable Future, 11

Brookline, Mass., 222, 223
Brown, Becky J., 4
Brown, David L., 7
Brownfield redevelopment, 12, 22, 32, 66, 78, 90, 91, 110
Brownfield sites in Jacksonville, 215
Brownfields Industrial Redevelopment Council (Baltimore), 92
Brownsville, Tex., 54, 78, 89, 115
Brugmann, Jeb, 61, 62, 68, 73, 60, 128, 197, 198, 243
Brundtland, Gro Harlem, 8
Brundtland Commission, 8, 10, 14, 16, 157, 159, 160, 232. *See also* United Nations' World Commission on Environment and Development
Bryant, Bunyan, 164
Bryner, Gary, 15
Bubble concept of sustainable cities, 18
Budgets, in sustainable cities, 74
Buffalo, N.Y., 147
Buga, Columbia, 21
Building Air Quality Alliance (San Francisco), 210
Building Livable Communities, 13
Bullard, Robert, 158, 163, 164
Burlington, Vt., 64
Business leaders, and sustainable cities, 109, 244–246

California Community Indicators Conference, 212
California, comprehensive planning requirements, 106
California, growth planning in, 109
California, State of, 25
Calthorpe, Peter, 19, 20
Camacho, David E., 164
Cambridge, Mass., 9, 22, 69, 94, 152, 153, 156, 222, 223, 224
Cambridge (Mass.) Civic Forum, 23, 50, 153, 223
Cambridge Growth Management Advisory Committee, 224

Campaign for Sustainable
    Milwaukee, 23
Canada, 21
Cancer alley (Louisiana), 164
Cape Town, South Africa, 21
Carbon dioxide emissions, in
    Portland, 208
Carley, Michael, 21
Carrying capacity, 4. *See also*
    Ecological carrying capacity
Cassidy, Robert, 13
Central Texas Indicators Project, 56,
    87
Chamber of Commerce in Santa
    Monica, 198
Chamber of Commerce in
    Jacksonville. *See* Jacksonville
    Chamber of Commerce
Charettes, in sustainable cities
    visioning processes, 249
Charles, John A., 106, 113, 209
Charlotte, N. C., 208
Chattanooga, Tenn., 22, 25, 29, 32,
    36, 54, 63, 69, 77, 79, 82, 83, 84,
    89, 91, 93, 99, 110, 116, 117,
    118, 155, 178, 179, 181, 185–193,
    216, 217, 219, 243
Chattanooga, 2020 Plan in, 190
Chattanooga Area Chamber of
    Commerce, 25, 110, 116, 118,
    186, 187, 188, 192, 244
Chattanooga Area Regional Transit
    Authority (CARTA), 99, 110
Chattanooga Creek, 84
Chattanooga Creek Cleanup Project,
    83–84
Chattanooga Greenways, 188
Chattanooga Hamilton County
    Regional Planning Agency
    (CHCRPA), 189, 191
Chattanooga Neighborhood
    Enterprise, Inc., (CNE), 188
Chattanooga Regional Growth
    Initiative, 189
Chattanooga Riverwalk, 118
Chattanooga Venture, 186

Chattanooga-Hamilton County
    Air Pollution Control Bureau,
    25
Chattanooga/Hamilton County
    Brownfields Program, 93
Chautauqua Conference on Regional
    Governance, 192
Chavis, Benjamin, 163
Chesapeake Bay, 91
Chicago, Ill., 89
Civic engagement, 147
Civic environmentalism, 133
Civic participation, 156
Civil society, 55, 56, 125, 126, 132,
    144
Cleveland, Ohio, 22, 78, 222,
    226
Cleveland EcoVillage, 226
Cleveland Neighborhood
    Development Corporation, 226
Clinton administration, 11, 13
Closed-loop definition of sustainable
    cities, 18
Cluster economic development, 67,
    110, 189
Coastal zone management, 12
Cobb, John, 137, 138, 139
Collaton, Elizabeth, 105
Colorado Noxious Weed Act, 203
Common-pool resources, 5, 134,
    136. *See also* Tragedy of the
    commons
Communalism, 136
Communitarian elements found in
    sustainable cities, 148–153
Communitarian foundations of
    sustainable cities, 125–156
Communitarian goals, 37
Community, conceptions of, and
    sustainability, 3, 11, 13
Community-based environmental
    protection, 12
Community Environmental Council,
    Inc., 44
Community and management of
    common-pool resources, 136

Community, rebuilding as a goal of sustainable cities, 145
Community Scorecard in Austin, 182
Comprehensive Emergency Response, Compensation, and Liability Act of 1980, 90
Comprehensive land use planning, 66
Comprehensive Plan, in Seattle, 194
Comprehensive planning, 20
Conservation, 32
Consortium for Regional Sustainability, Tufts University, 50
Cook, Fay Lomax, 147
Corson, Walter, 48, 49, 54
Crime rates as indicators of sustainability, 47
Critique-of-technology elements of sustainability, 4
Cronkite, Walter, 186
Cryptosporidium, 83
Cuyahoga County Planning Commission, 226
Cuyahoga River, 78

Dale, A., 21
Daly, Herman, 137, 138, 139
DeLeon, Richard E., 235
Deliberative democracy, 146
Democracy building, 59
  as a component of sustainable cities, 45
Demographic characteristics of sustainable cities, 179
Dental sector, in Boulder, 207
Denver, Colo., 122, 202, 208, 222
Des Moines, Iowa, 147
Developing indicators, the process of, 59–62
Development regime, 26. See also Urban regime
Diamond, Henry L., 19
Dioxin, 163
Diversity, 37
Dobson, Andrew, 160
Don't Top It Off (Boulder), 204

Dore, Mohammad, 164
Downs, Anthony, 122, 248, 249
Downtown business elite in urban governance regimes, 26. See also Urban regime
Dryzek, John, 17, 137
Dumping in Dixie, 163
Dunphy, Robert T., 19, 98
Durban, Johannesburg, 21
Duval County, Fla. (Jacksonville), 25, 212

Earth Summit, 10
EcoCity, Cleveland, 23, 226
EcoCycle, in Boulder, 205
Ecodevelopment elements of sustainability, 4, 8
Eco-Industrial Park (Brownsville, Tex.), 23
Eco-industrial parks, 12, 36, 54, 66, 89, 110, 115–116
Ecological carrying capacity, 4, 5, 7
Ecological footprint, 18, 19, 20, 21, 32, 34, 47, 49, 52, 102, 141, 161
Ecological footprint, and transboundary shifting, 143
Ecological integrity, as a dimension of sustainability, 50, 52
Eco-neighborhoods, 12
Economic efficiency vs. sustainability, 106
Economic equality, 37
Economic growth, traditional conceptions, 102
Economic security, 50, 105
Economic security, disparities, 161
Ecoracism, 158, 164
Ecosystem health, 9
Ecosystem indicators, 196
Ecosystems, 20
Ecosystems and cities, 14
Ecosystems and communities, 11
EcoVillage at Cleveland, 12
EcoVillage at Ithaca, 12, 23
Eco-villages, 12, 66

Electric Transit Vehicle Institute (ETVI) (Chattanooga), 82, 99, 110
Empower Baltimore Management Corporation, 92
Empowerment, 68
 as a goal in sustainable cities initiatives, 51, 127
Energy conservation, 67
Energy use and conservation, 95–97
English, Mary P., 109
Enterprise Community (Portland), 92, 93
Environmental Advisory Board, in Boulder, 203
Environmental equity, 57, 157–175
 defined, 162
Environmental justice, 37, 57, 157–175. *See also* Environmental equity
Environmental justice movement, 158
Environmental Management Program (Seattle), 195
Environmental management system (EMS), in Seattle, 195
Environmental overlay zones, in Portland, 208
Environmental Policy Center, 49
Environmental Protection Agency (EPA). *See* U.S. Environmental Protection Agency (U.S. EPA)
Environmental protection hustle, 167, 168
Environmental Protection Zones (Portland), 112
Environmental racism, 163. *See also* Environmental justice
Environmental Sustainability Index, 33
Equity
 in access to capital, 173
 in education, 172
 in law enforcement, 173
 in leadership positions, 173
Esty, Daniel C., 33, 229
Ethnic status and environmental justice, 162

Etzioni, Amatai, 132
Evans, Bob, 134

Fainstein, N. I., 26
Fainstein, S. S., 26
Fair share proposals and policies, 166
Federalism, 15
Ferguson, L. Joe, 186
Feyerabend, Grazia Borrini, 13
Fischel, W. A., 106
Florida, comprehensive planning requirements, 106
Florida, State of, 249
Foreman, Christopher H., 164
Fox, Jonathan A., 7
Fox-Wolf Basin (Wisconsin), 12
Framework for community-based environmental protection, 12
Franciosi, Robert, 106
Francis, George, 6
Free Trade Agreement of the Americas, 15
Free-ridership, 139
Frieden, Bernard J., 108, 167, 168
Fulton, William, 19, 20
Futurescape Community Planning Process, 189, 190

Garbage imperialism, 164
Gardner, Roy, 137
Geiser, K., 163
Georgia, State of, 109, 185
Global Cities Online Project, 37, 49, 169
Global Tomorrow Coalition, 49
Global warming, program to combat, in Portland, 207–208
Goldman, Benjamin A., 164
Goode, Ann Eberhart, 105
Gordon, Peter, 106
Governance mechanisms for sustainability, 16, 68
Governance regimes, 245
Graham, Lamar, 186
Grantsville, Utah, 64, 239

Grantsville (Utah) General Plan for Sustainable Community, 23
Graphic Packaging, Inc., 203
Great Lakes Basin, 12
Green backlash, 130
Green Building Initiative (Portland), 209
Green Building program (Austin), 184
Green building programs, 67, 96, 119–121
Green Index, 33
Green Metro Index, 33
Green Points initiative (Boulder), 97, 120, 203, 204
GreenChoice (Austin), 97, 184
Greenhouse gases, 80
Greenlight Foundation, 89
Greenway zones, 112
Greenways program (Boulder), 203
Groundwater Resources Protection Program (Chattanooga), 83
Growth boundaries, 20
Growth machine, conceptions of in cities, 17, 102, 103, 105
Guy, Simon, 16, 17

Hall, Bob, 33
Hamilton County, Tenn., 25, 84, 87
Hamilton, Canada, 238
Hamilton, New Zealand, 21
Hamilton-Wentworth, Canada, 21
Hanson, Mark E., 4
Hardin, Garrett, 134
Hart, S., 98
Hartford, Conn., 143
Haughton, G., 162
Hauser Chemical Research, Inc., 203
Hays County, Tex., 25
Hazardous materials management, 32
Hazardous waste, 20, 88
Hazardous waste recycling, 67
Hazardous waste sites, 53
Hillsborough County, Fla., 227
Historic preservation, 27
Hogen-Esch, Thomas, 115

Household recycling, 22
Household recycling programs, 32
Household solid waste recycling, 67. *See also* Recycling solid waste
Houston, Tex., 164
Hybrid-electric vehicles, 110

ICI Americas (Chattanooga), 116
ICMA. *See* International City/County Management Association
Iglitzin, Lynne B., 114
Index of Taking Sustainability Seriously, correlates of, 229–238
Index of Taking Sustainability Seriously, elements of, 64–72
Indicators
  of agriculture, 41
  of air quality, 41
  of biodiversity, 41
  of biodiversity, in San Francisco, 211
  of climate change, 41
  of cultural issues, 49
  of ecological health, 52
  of economic development, 41
  of education, 41
  of energy, 41
  of energy consumption, 53
  of environment, 52
  of environmental justice, 41
  of equity and equality, 57
  of ethical issues, 49
  of food, 41
  of governmental function, 49
  of hazardous material, 41
  of human health, 41
  of local economic performance, 53
  of ozone depletion, 41
  of parks and open spaces, 41
  of resource conservation (Santa Monica), 199
  of social issues, 49
  of sustainability, 36–59
Indicators of Community Well-Being (Chattanooga-Hamilton County, Tenn.), 25

Indicators projects, 41, 62
Industrial or commercial solid waste
  recycling, 67
Industrial pollution, 88
Industrial pollution, disparities in
  exposure to, 162
IndyEcology, 23
Inequality, 157. *See also*
  Environmental equity
Ingram, Helen, 143
International City/County
  Management Association (ICMA),
  66, 105
International Council for Local
  Environmental Initiatives (ICLEI),
  21
International Monetary Fund (IMF),
  7
Ithaca, N.Y., 64, 239

Jacksonville, Fla., 22, 25, 29, 40, 44,
  56, 57, 63, 69, 86, 87, 117, 150,
  151, 155, 169, 171, 178, 179,
  212–216, 217, 219, 243
Jacksonville Chamber of Commerce,
  171, 213, 214
Jacksonville Community Council Inc.,
  23, 56, 150, 171, 213, 214, 217
Jacksonville Comprehensive Plan,
  215
Jacksonville Equal Opportunity
  Commission, 171
Jacksonville Indicators Project, 23
Jacobs, Lawrence R., 147
Jakarta, Indonesia, 160
Jinja, Uganda, 21
John, DeWitt, 133, 134, 148
Johnson, Curtis, 189, 192
Johnstone Shire, Australia, 21
Judd, Dennis R., 103

Kansas City, Mo., 208
Kee, Warner, 168
Kenworthy, Jeffrey, 67, 98, 99
Kerr, Mary Lee, 33
Kidd, Charles V., 4, 5, 8, 9

King County, Wash., 148
Kjellen, Marianne, 160, 161
Kline, Elizabeth, 44, 48, 50, 55, 59,
  61, 105, 121, 151, 161, 162, 169
Kraft, Michael, 12

Lachman, Beth E., 12, 67
Lake Michigan, 83
Lansing/East Lansing, Mich., 64
Lawn Care and Integrated Pest
  Management (Boulder), 204
Lead paint abatement, 67
Lead paint, disparities in exposure to,
  162
Leadership in Energy and
  Environmental Design rating
  system, 209
Ledger, Marshall A., 147
Lee, Charles, 163
Legge, Russel, 6
Leo, Christopher, 26, 113
Lerner, Sally, 6
Lewis, Paul, 168
Lexmark International, Inc., 203
Liberalism, 132. *See also* Political
  liberalism
Linkage funds, 27
Liroff, Richard E., 78
Livability, 28
Livable cities, 1, 13
Livable Tucson Vision Program, 23
Liverman, Diana M., 4
Living Laboratory (Chattanooga),
  99
Local Agenda 21, 134
Local Agenda 21 Model
  Communities Programme, 21
Local authorities' initiatives in
  support of Agenda 21, 10
Local commerce, 121
Longmont, Colo., 81, 117
Lookout Creek, Chattanooga, 188
Lookout Mountain, Chattanooga,
  188
Los Angeles, Calif., 79, 197
Loveridge, Ronald O., 168

Lower class opportunity expansion regime, 27. *See also* Urban regime
Lyndhurst Foundation, 186

MacKenzie, W. R., 83
MacLaren, V. W., 61
Magilavy, Beryl, 212
Maintenance regime, 26. *See also* Urban regime
Management of recreational waterways, 79
Mandershield, Melissa, 109
Manning, Kristy, 1, 12, 18, 37, 118
Manufacturing industries and local politics, 16
Marvin, Simon, 16, 17
Maryland, growth planning in, 109
Maser, Elisa, 214
Mass transit, 67. *See also* Public transit
Matheny, Albert R., 148
Mauritania, 13
Maximum carrying capacity, 5. *See also* Ecological carrying capacity
Mazmanian, Daniel, 12, 148
McGranahan, Gordon, 160, 161
Measurement of sustainable cities, 28
Measurement of Taking Sustainability Seriously, 240. *See also* Index of Taking Sustainability Seriously
Menino, Thomas, 154
Meredith, Robert W., 4
Metropolitan areas and sustainable cities, 24
Metropolitan Council for Community Services (Chattanooga), 187
Metropolitan Milwaukee Sewerage District, 83
Metropolitan-wide planning, 20
Middle class progressive regime, 27, 104, 144. *See also* Urban regime
Milbrath, Lester, 16
Milich, Lenard, 143
Millenium [sic] Project (Olympia, Wash.), 40

Milwaukee, Wis., 69, 83
Minneapolis, Minn., 89, 208
Mlay, M., 134
Moe, Richard, 19
Mohai, Paul, 164
Molotch, Harvey, 17, 103, 168
Moore, Terry, 109, 112
Mount, Timothy, 164
Myers, Phyllis, 107

NAACP, 163
National Commission on Civic Renewal, 131, 145
National Commission on the Environment (NCE), 8
National Environmental Technology Strategy, 11
National Priorities List (NPL), 84, 86, 90. *See also* Superfund
National Science and Technology Council (NSTC), 11, 12
Natural capital, 8, 9
Natural resource/environment elements of sustainability, 4
Neiman, Max, 168
Nelson, Arthur C., 109, 112
Neoclassical economics, 130, 138 and the tragedy of the commons, 135
Nepal, 12
New Haven, Conn., 22, 24, 222
New localism of environmental policy, 16
New Property Rights movement, 130
New York (city), 19
Newby, Howard, 11
Newman, Peter, 67, 98, 99
NeXstar Pharmaceuticals, Inc., 204
NIMBY opposition, 166
NIMBY Syndrome, 125, 129, 131, 139–142, 156, 250 and environmental equity, 165
NIMTOO (not in my term of office), 141
No Drive Days (Boulder), 204

No-growth sentiment, 9
Nonprofit organizations and
  sustainable cities initiatives, 40
Nonprofit sector, 22, 35, 39, 167
Noonan, Patrick E., 19
North American Free Trade
  Agreement (NAFTA), 15
North Chickamauga Creek
  (Chattanooga), 84, 188
North Chickamaugua Creek Gorge
  Watershed Protection Project
  (Chattanooga), 83
Northeast Ohio Environmental
  Monitoring for Public Access and
  Community Tracking (NEO-
  EMPACT), 226

Oakland, Calif., 115
Odum, E. P., 5
Oklahoma City, Okla., 64
Olympia, Wash., 25, 40, 78, 121,
  154, 222, 230, 249
Olympia, Wash., Sustainability
  Roundtable, 63
Open space planning, 20
Ophuls, William, 94, 128, 135, 136,
  144
Optimum carrying capacity, 5. *See
  also* Ecological carrying capacity
Oregon, city park ballot question in,
  107
Oregon, growth planning in, 109
Orlando, Fla., 69
Ostrom, Elinor, 136, 137, 145

PACE. *See* Partners for a Clean
  Environment
Pallemaerts, Marc, 143
Pallisa Community Development
  Trust (Uganda), 12
Parker, Jane, 134
Parr, John, 190, 192, 186
Participation indicators, 149
Partners for a Clean Environment
  (PACE), in Boulder, 81, 88, 117,
  118, 203, 206, 207, 244

Partners for Economic Progress
  (Chattanooga), 189
Partners for Livable Communities,
  13
Partnership for Sustainable Brookline
  Project, 225
PCBs, 163
Peiser, Richard B., 106
Peretz, Jean H., 109
Perkins, Jane, 164
Pew Charitable Trusts, 147
Phoenix, Arizona, 122, 147, 208,
  230
Pierce, J. T., 21
Pierce, Neal, 189, 190, 192
Pinfield, Graham, 73
Pittsburgh, Pa., 54, 89
Plattsburgh, N.Y., 115
Political liberalism, 130. *See also*
  Liberalism
Political machines, 15
Politics of sustainable cities, 153–155
Pollution prevention, 81
Population characteristics of
  sustainable cities, 180
Population density, 25
  as a correlate of sustainability, 217
Population growth, as a correlate of
  sustainability, 217
Portland, Oreg., 22, 29, 36, 40, 66,
  69, 72, 73, 91, 92, 99, 107, 110,
  112, 178, 179, 181, 190, 207–210,
  217, 219, 228, 235
Portland Brownfields Initiative, 92
Portland Bureau of Environmental
  Services, 208
Portland Comprehensive Plan, 209
Portland Office of Sustainable
  Development, 63, 208
Portney, Kent E., 139, 147, 163, 166
Possibilities: Neighbors in Action
  (Oklahoma City, Okla.), 23
Potapchuk, William, 134
Presidential elections, 234
  voting in, as a correlate of
  sustainability, 218, 219

Printing sector, in Boulder, 207
Private sector involvement, 117–119
Public agenda, 32
Public participation, 37
  in Chattanooga, 192
  in sustainable cities initiatives, 127
Public transit, 67, 98–99. *See also*
  Mass transit
Puentes, Robert, 107
Puget Sound, 193
Putnam, Robert, 55, 131
Pye-Smith, Charlie, 13

Quality of life and environmental
  justice, 158
Quality of life indicators, 28, 39,
  50
  in Jacksonville, 171
Quality of Life indicators project
  (Jacksonville), 44, 150, 171, 213

Rabe, Barry G., 148
Racial and environmental justice,
  162. *See also* Environmental equity
Radical decentralization of
  environmental decisions, 17, 137
Radioactive colonialism, 164
Rampant individualism, 136
  as a cause of unsustainability, 129
Recycle Boulder, 205
Recycling solid waste, 87–88. *See
  also* Solid waste
Redefining Progress (Oakland, Calif.),
  196, 212
Rees, Mathew, 163
Rees, William E., 18, 19, 20, 34, 49
Referenda to regulate urban growth,
  107
Renewable energy, in Austin, 184
Renewable energy sources, 67, 68,
  80
Restaurant sector, in Boulder, 207
ReVision 2000 (Chattanooga), 186,
  188, 190
Rice, Norm, 114
Richardson, Harry W., 106

RiverCity Company (Chattanooga),
  189
RiverValley Partners (Chattanooga),
  118, 189
Riverwalk (Chattanooga), 189
Robinson, John, 6, 161
Roche Colorado, Inc., 204
Rosenbloom, Sandra, 19
Russ, Thomas H., 90

Sachs, Ignacy, 8
Sacramento, Calif., 208
San Francisco, Calif., 22, 25, 29, 57,
  68, 69, 72, 83, 85, 86, 94, 96,
  115, 119, 155, 169, 170, 171, 175,
  178, 179, 181, 210–212, 217, 219,
  235
  Sustainability Plan, 94
San Francisco Bay Area, 167
San Francisco Commission on the
  Environment, 211
San Francisco Sustainable City, 87
San Francisco sustainable indicators,
  42–44
San Jose, Calif., 22, 25, 69, 96, 178,
  222, 228, 235
San Jose 2020, 23, 228
Sandel, Michael J., 132
Santa Barbara, Calif., 22, 40, 44, 56,
  57, 150, 222, 235
Santa Monica, Calif., 9, 22, 25, 29,
  40, 68, 69, 72, 79, 80, 95, 96,
  119, 175, 178, 179, 181, 190,
  197–202, 216–217, 235
  sustainability indicators, 200
Santa Monica Civic Center, 96
Santa Monica Department of
  Environmental and Public Works
  Management, 199
Santa Monica Municipal Bus Line,
  201
Santa Monica Pier, 96
Santa Monica Sustainable City
  Program, 23
Santos, Brazil, 21
São Paulo, Brazil, 160

Satterthwaite, David, 17
Saunders, Sid, 16
Scandinavia, 21
Scottsdale, Ariz., 22, 53, 69, 119,
  178, 222
Scottsdale Seeks Sustainability, 23
Seattle, Wash., 22, 25, 29, 36, 40,
  41, 56, 60, 66, 69, 72, 78, 79, 80,
  94, 148, 147, 150, 152, 156, 169,
  170, 179, 190, 193–197, 208, 209,
  216, 219, 222, 228, 235, 249
Seattle Comprehensive Plan, 55, 58,
  80, 81
Seattle Comprehensive Plan Land Use
  Element, 114
Seattle Comprehensive Planning,
  149
Seattle Office of Sustainability and
  Environment, 63
Second Livestock Project in
  Mauritania, 13
Selman, Paul, 16, 134
Shaiko, Ronald G., 15
Shared community values, 136
Silverstein, Judith, 214
Singsore, Jacob, 160, 161
Skinnarland, Kirvil, 195
Smart development, 78. *See also*
  Smart growth
Smart growth, 28, 54, 66, 78, 89,
  101–123, 104–108, 105, 110,
  174
critiques of, 122
Smart Growth Network, 109, 115,
  116
Smith, D. Clayton, 141, 163
Social capital, 249
Social and environmental justice, 57.
  See also Environmental equity
Social equity, 37. *See also*
  Environmental equity
Social justice, 157. *See also*
  Environmental equity
Social Security Reform, 147
SolarSource energy program
  (Boulder), 204

Solid and hazardous waste
  management, 86–94
Solid waste, 19, 20
and recycling indicators, 41
transfer station, 165
Sound Exchange (Olympia, Wash.),
  121, 223
South Central Business District Eco-
  Industrial Park (Chattanooga), 116
South Chattanooga, 188
South Chickamauga Creek
  (Chattanooga), 188
South Coast Community Indicators
  Project (Santa Barbara), 23, 150
Spapens, Philippe, 21
Spivak, A. L., 98
Spohn, Suzanne G., 116
Sprawl, 25, 107, 248. *See also* Urban
  sprawl
St. Paul, Minn., 208
Stakeholders, participation of in
  sustainability initiatives, 60, 153
Staten Island, N.Y., 19
Steady-state society, 136
Stone, Clarence, 26, 27, 104, 144,
  245
Strategic planning and sustainability,
  45
Strategic plans and sustainable cities
  initiatives, 36
Stuart, Fla., 64, 239
Superfund, 53, 90. *See also* National
  Priorities List
Superfund site remediation, 90
component of sustainability
  initiatives, 67
and sustainable cities initiatives, 86
Superfund sites, in Chattanooga, 84
Superfund sites, disparities in
  proximity to, 163
Sustainability indicators, 38, 44, 45,
  66
Sustainability Indicators Project of
  Hays, Travis, and Williamson
  Counties (Austin), 23, 172, 173,
  174

Sustainability Indicators Project of
Hays, Travis, Williamson,
Caldwell, and Bastrop Counties
(Austin), 181
Sustainability indicators projects, 37,
40
Sustainability initiatives, 32, 35
Sustainability Plan (San Francisco),
210
Sustainability plan, as a component
of taking sustainability seriously,
36
Sustainability Roundtable (Olympia,
Wash.), 63, 154, 223. See also
Sustainable Community
Roundtable
Sustainability 2000 Project (Boulder),
203, 205
Sustainable agriculture, 4
Sustainable biological resource use, 4
Sustainable Boston Initiative, 23, 224
Sustainable Chattanooga, 23, 36, 84.
See also Chattanooga, Tenn.
Sustainable cities, 1
Sustainable City San Francisco, 171,
211, 217
Sustainable City Major Strategy (San
Jose), 23, 228
Sustainable City Program in Santa
Monica, 198
Sustainable City Progress Report
(Santa Monica), 201
Sustainable City Working Group,
Santa Monica, 199
Sustainable Cleveland Partnership,
23, 226
Sustainable communities, 1
Sustainable Communities Advisory
Board, Tampa, 228
Sustainable communities indicators,
48. See also Sustainability
indicators
Sustainable Communities program in
Orlando, Fla., 23
Sustainable Community Roundtable
(Olympia, Wash.), 23, 40, 66, 121.

See also Sustainability Roundtable
Sustainable development, 4, 7–9
Sustainable economic development, 3,
7. See also Sustainable development
Sustainable Energy Task Force
(Austin), 184
Sustainable energy, 4
Sustainable governance, 55
Sustainable indicators initiatives, 28.
See also Sustainability indicators
Sustainable Indicators Project in
Seattle, 194
Sustainable Lansing, 23
Sustainable Portland Commission,
208
Sustainable Seattle indicators, 46–
47
Sustainable Seattle model, 196
Sustainable Seattle, 23, 36, 49, 60,
62, 80, 194, 152, 169, 170, 193,
196, 213, 217, 223, 249
Sustainable society and economy, 4
Swanstrom, Todd, 103
Switzer, Jacqueline Vaughn, 130

Taking Sustainability Seriously Index,
64–72, 230
Tampa, Fla., 22, 40, 41, 178, 222,
227
Tampa/Hillsborough Sustainable
Communities Demonstration
Project, 23
Target for 2000 (Jacksonville), 213
Target for 2005 (Jacksonville), 213
Targeted economic development, 66,
110
Tax incentives, 66
Tennessee, State of, 185
Tennessee American Water Company,
85
Tennessee Aquarium, 118, 189
Tennessee Department of
Environment and Conservation,
84
Tennessee, growth planning in,
109

Tennessee River, 84
Tennessee River Gorge Project, 83
Tennessee River Watershed, 83
Tennessee Riverpark, Chattanooga, 188, 189
Tennessee Riverpark Project, 83
Tennessee Valley Authority, 25, 84, 99
Thomson, Ken, 147
Toronto, Canada, 21, 222, 238
"Toward a Sustainable City" (San Jose), 228
"Toward a Sustainable Seattle," 36, 152, 194
"Towards a Sustainable Future" (Cambridge), 223
Toxics Release Inventory, in Austin, 182
Tragedy of the commons, 125, 129, 131, 134–139. *See also* Common-pool resources
Transboundary environmental impacts, 78, 125, 131, 142–144
Transit-oriented development in Austin, 185
Transportation planning, 98–99. *See also* Mass transit
Trash incinerators, location of, 165
Travis County, Tex., 25
Traynor, Michael, 164
Trust for Public Land, 188
Tucson, Ariz., 115
Tufts University, 50

Uganda, 12
United Church of Christ, (UCC), 163, 164
United Nations' World Commission on Environment and Development (WCED), 8, 10, 14, 16, 159, 160. *See also* Brundtland Commission
United Nations Environmental Programme, 8, 10
U.S. Agency for International Development (USAID), 7
U.S. Bureau of the Census, 41, 74

U.S. Bureau of Labor Statistics, 41
U.S. Environmental Protection Agency (U.S. EPA), 12, 41, 84, 90, 91, 164, 215
U.S. Government Accounting Office, 163
U.S. Green Building Council, 209
United Way of Chattanooga, 188
United Way in Jacksonville, 214
United Way in Santa Barbara, 150
University of Tennessee at Chattanooga, 84
Urahn, Susan K., 147
Urban footprint, 18. *See also* Ecological footprint
Urban regime, 24, 26
Urban sprawl, 24, 104. *See also* Sprawl
Urban villages, 114, 195. *See also* Eco-villages
USEPA. *See* U.S. Environmental Protection Agency (U.S. EPA)

Vancouver, Canada, 238
Varady, Robert G., 143
Virginia, State of, 19
Vision for a Greater New Haven, 23
Vision 2000 (Chattanooga), 186, 188, 190
Visioning process, 41, 68, 249
Visioning process, in Chattanooga, 186
Visual Preference Survey in Chattanooga, 190
Voice of Customer Survey in Austin, 182
Volatile organic compounds (VOCs), 82
Volunteer Site (Chattanooga), 116

Wackernagel, Mathis, 18, 34
Walker, James, 137
Walker, Natalie, 164
Walsh, Edward J., 141, 163

Waneck, G., 163
Warland, Rex, 141, 163
Warren County, N. C., 163
Washington, State of, 25
  comprehensive planning
    requirements, 106, 109
  Growth Management Act, 114
Waste, Robert, 14
Wastewater treatment, 53, 79
Water management, 32, 78
Water quality, 82–86
Watershed management, 79
WCED. *See* United Nations' World
  Commission on Environment and
  Development
West Seattle, 115
Western Europe, 21
Wetlands protection, 79
Wilkie, Carter, 19
Willamette River, 92
Williams, Bruce A., 148
Williamson County, Tex., 25
WindSource energy program
  (Boulder), 204
Winnipeg, Canada, 238
World Bank, 7
World Resources Institute, 33

Young, Dwight, 104
Young, Oran, 143

Zachary, Jill, 9, 44, 45, 60, 61, 66,
  81, 151, 152
Zero-emission buses, 189
Zoning, 66, 110
  and comprehensive land use
    planning, 111–115
  for the environment, 112